IMPACT ASSESSMENT

of

HAZARDOUS AQUATIC CONTAMINANTS

Concepts and Approaches

Edited by

Salem S. Rao

National Water Research Institute,
Canada Centre for Inland Waters,
Burlington, Ontario, Canada

LEWIS PUBLISHERS
Boca Raton London New York Washington, D.C.

Acquiring Editor:	Skip DeWall
Project Editor:	Sylvia Wood
Marketing Managers:	Barbara Glunn and Jane Stark
Manufacturing Manager:	Carol Slatter

Library of Congress Cataloging-in-Publication Data

Rao, S.S. (Salem S.), 1934—
    Impact assessment of hazardous aquatic contaminants: concepts and
approaches / Salem S. Rao
        p. cm.
        Includes bibliographical references and index.
        ISBN 0-8493-4109-4
        1. Water—Pollution—Environmental aspects 2. Aquatic organisms—
Effect of water pollution on. 3. Ecological risk assessment.   I. Title.
    QH545.W3 R26 1999
    577. 6′27—dc21
    DNLM/DLC
    for Library of Congress   98-47582
                                                                      CIP

© 1999 by CRC Press LLC
Lewis Publishers is an imprint of CRC Press

No claim to original U.S. Government works
International Standard Book Number 0-8493-4109-4
Library of Congress Card Number 98-47582
Printed in the United States of America 1 2 3 4 5 6 7 8 9 0
Printed on acid-free paper

Acknowledgments

The Editor would like to thank all contributors who made this important and timely volume possible. In particular, he would like to thank Dr. John Carey, Executive Director, N.W.R.I., and Dr. K.R. Munkittrick for their encouragement and support, and Dr. B.K. Burnison, for his computer assistance during this project.

About the Editor

Salem S. Rao, Ph.D., is a Scientist Emeritus at the National Water Research Institute, Canada Centre for Inland Waters, Burlington, Ontario, Canada. His studies have contributed largely to the area of environmental microbiology and toxicology. He is the Editor/Author of *Acid Stress and Aquatic Microbial Interactions* and *Particulate Matter and Aquatic Contaminants*. Dr. Rao has worked in the area of Aquatic and Environmental Microbiology and Ecotoxicology for over 30 years. His current research interest lies in the area of contaminant impact assessment on aquatic biota. During his professional career he has published over 100 research papers and reports on such diverse topics as bacterial-nutrient dynamics in the Great Lakes and in acid stressed lakes; impact of acid stress on aquatic microbes, particulate matter, and aquatic contaminant interactions; and impact of environmental toxins, genotoxins, and mutagens on fish.

Introduction

The Canadian Environmental Protection Act (CEPA) is an act which deals with the protection of the environment and human health, with provisions for dealing with toxic substances and nutrients, and environmental effects operations.

During the last few years there has been increased activity in the area of environmental effects monitoring, commonly known in Canada as EEM. These EEM programs are generally used to assess the effectiveness of environmental regulatory programs and mitigative measures to actually protect aquatic environments. Consequently, the number and types of approaches in environmental effects monitoring programs are rapidly expanding.

The impetus behind the compilation of this volume is the ever-increasing awareness of the importance of environmental contaminant impact assessment on aquatic environments. The impact of environmental contaminants (toxins and genotoxins) on biota in aquatic environments is fairly well recognized. However, there are still some unknowns regarding these adverse effects assessment approaches on aquatic organisms at different levels of organization (subcellular, cellular, and organism level). It has been surmised that a fuller understanding of the effects of these contaminants on aquatic biota at these levels would enhance our understanding of long-term effects, and better contribute to aquatic management strategies.

This volume presents information and techniques which are at the leading edge of environmental toxicology. The eight chapters in this volume are timely, and address a number of key environmental issues of global concern. Hence, the concepts and approaches used in environmental impact assessment studies are of special interest to environmental researchers and managers of aquatic toxicology research programs.

Contributors

Christian Blaise
Bioanalytical Research Unit
Centre Saint-Laurent
Environment Canada
105 McGill Street
Montreal, Quebec
Canada H2Y 2E7

B. Blunt
National Water
    Research Institute
Canada Centre for
    Inland Waters
867 Lakeshore Road
Burlington, Ontario
Canada L7R 4A6

B. Kent Burnison
National Water Research
    Institute
Canada Centre for
    Inland Waters
867 Lakeshore Road
Burlington, Ontario
Canada L7R 4A6

Richard Chong-Kit
Biohazard Laboratories
Aquatic Toxicology Section
Ontario Ministry of the
    Environment and Energy
125 Resources Road
Etobicoke, Ontario
Canada M9P 3V6

A. Gamble
Department of Environmental
    Biology
University of Guelph
Guelph, Ontario
Canada N1G 2W1

K. Gorman
GLLFAS, Department of Fisheries
    and Oceans
Canada Centre for Inland Waters
867 Lakeshore Road
Burlington, Ontario
Canada L7R 4A6

Michelle A. Gray
Watershed Ecosystems
    Program
Trent University
Peterborough, Ontario
Canada K9J 7B8

P. Hansen
Department of Ecotoxicology
Institute for Ecology
Technical University of Berlin
Berlin, D-10589
Germany

M. Anthony Hayes
Department of Pathobiology
Ontario Veterinary College
University of Guelph
Guelph, Ontario
Canada N1G 2W1

B. Hock
Department of Botany
Technical University of Munich
Weihenstephan
Freising, 85350
Germany

P. Hodson
Department of Environmental
   Biology
Queens University
Kingston, Ontario
Canada K7L 3N6

Gordon M. Kirby
Department of Biomedical
   Sciences
University of Guelph
Guelph, Ontario
Canada N1G 2W1

Yiannis Kiparissis
Watershed Ecosystems
   Program
Trent University
Peterborough, Ontario
Canada K9J 7B8

Takashi Kusui
College of Technology
Toyama Prefectural University
5180 Kurokawa, Kosugi-Machi
Imizu-Gun, Toyama 939-03
Japan

A. Marx
Department of Botany
Technical University of Munich
Weihenstephan
Freising, 85350
Germany

M.E. McMaster
National Water Research Institute
Canada Centre for Inland Waters
867 Lakeshore Road
Burlington, Ontario
Canada L7R 4A6

C. D. Metcalfe
Environmental and Resource Studies
Trent University
Peterborough, Ontario
Canada K9J 7B8

K.R. Munkittrick
National Water Research Institute
Canada Centre for Inland Waters
867 Lakeshore Road
Burlington, Ontario
Canada L7R 4A6

Joanne L. Parrott
National Water Research Institute
Canada Centre for Inland Waters
867 Lakeshore Road
Burlington, Ontario
Canada L7R 4A6

Salem S. Rao
National Water Research Institute
Canada Centre for Inland Waters
867 Lakeshore Road
Burlington, Ontario
Canada L7R 4A6

David A. Rokosh
Biohazard Laboratories
Aquatic Toxicology Section
Ontario Ministry of the
   Environment and Energy
125 Resources Road
Etobicoke, Ontario
Canada M9P 3V6

Alan G. Seech
Grace Bioremediation Technologies
3465 Semenyk Court, 2nd Floor
Mississauga, Ontario
Canada L5C 4P9

M.R. Servos
National Water Research Institute
Canada Centre for Inland Waters
867 Lakeshore Road
Burlington, Ontario
Canada L7R 4A6

J. Sherry
National Water Research Institute
Canada Centre for Inland Waters
867 Lakeshore Road
Burlington, Ontario
Canada L7R 4A6

K. Solomon
Centre for Toxicology
University of Guelph
Guelph, Ontario
Canada N1G 2W1

Jack T. Trevors
Department of Environmental
    Biology
University of Guelph
Guelph, Ontario
Canada, N1G 2W1

G.J. Van Der Kraak
Department of Zoology
University of Guelph
Guelph, Ontario
Canada N1G 2W1

Table of Contents

Chapter 1

Neoplastic and Inflammatory Liver Diseases of White Suckers as Environmental Quality Indicators

M.A. Hayes and G.M. Kirby

INTRODUCTION

White suckers *(Catostomus commersoni)* are bottom-feeding fish native to fresh-water habitats in North America. Liver neoplasms are more frequently observed in white suckers captured from the more polluted urban/industrial regions of western Lake Ontario than in those from less polluted reference sites in the Great Lakes.[1-5] Liver neoplasms of white suckers in these locations are associated with an endemic chronic inflammatory liver disease which is more widespread and present in reference sites in which liver tumors are rare.[3,6,7] Liver neoplasms have also been observed in brown bullheads in some polluted locations in Lake Ontario[3,5] and Lake Erie.[8,9] Such neoplastic and other chronic diseases are potential indicators of impacts of environmental factors that are potentially harmful to health and longevity of resident fish populations. Indirectly, these conditions might also be bioindicators of other adverse environmental changes within the fish habitats, or remotely at the sources of contamination.

Increased prevalence of neoplasms in bottom-feeding fish has frequently been proposed as an indicator of environmental impact by carcinogenic contaminants.[10-12] Various studies have demonstrated correlations between sediment contamination and liver neoplasms in various marine[13-16] and freshwater locations.[2,3,5,17] However, these approaches have not led to rapid identification of particular environmental carcinogens, in part because of the many environmental contaminants and other influences that might be risk factors for neoplastic development. Also, neoplastic development in vertebrates usually involves multiple pathogenetic steps and multiple etiological factors[18,19] so retrospective interpretation of epidemiological data from field surveys is not a routine undertaking. This chapter describes various liver diseases of white suckers in the Great Lakes, and considers aspects of their pathogenesis which influence the monitoring and interpretation of these conditions as measures of environmental impacts of hazardous aquatic contaminants.

LIVER DISEASES OF WHITE SUCKERS

The salient features of some of the major neoplastic and inflammatory liver diseases of white suckers from Lake Ontario have been described previously.[3,6,20–22] There are various similarities among the types of neoplasms observed in white suckers and those described in brown bullheads *(Ictalurus nebulosus)* in the Great Lakes[3,5,8,11] and also in English sole in Puget Sound.[13,14,23] Experimentally generated liver neoplasms of fish exposed to carcinogens also resemble the hepatocellular preneoplastic and neoplastic lesions of white suckers, but experimentally exposed fish lack the chronic inflammatory lesions observed in wild fish.[24–26] White suckers throughout the Great Lakes also have various benign neoplastic skin papillomas and plaques,[2,3,27] but these are not considered here. These skin tumors are widespread in the Great Lakes populations of white suckers and more frequent in more contaminated regions, but occur at high frequencies in fish from cleaner reference sites.

Liver Neoplasms

Liver neoplasms and preneoplastic lesions observed in white suckers are mostly of bile duct or hepatocellular types and at various stages of progression to advanced aggressive malignancy. Hepatocellular lesions include phenotypically altered foci of preneoplastic hepatocytes of various basophilic, eosinophilic, or clear cell types (Figure 1.1). The altered hepatocellular foci are considered preneoplastic, but since they are well-recognized stages of hepatocarcinogenesis, they are included here with the later stages of neoplasia. Similar larger nodular proliferations of altered hepatocytes are termed hepatocellular adenomas (Figure 1.2) or hepatomas if they are large but nonaggressive. These may be grossly visible as round discolored nodular lesions within the hepatic parenchyma (Figure 1.3). Larger, locally aggressive hepatocellular carcinomas are more pleomorphic and multinodular in appearance; some have well-defined hepatocyte morphology, whereas others are more cystic and tubular in microscopic appearance (Figure 1.4). These carcinomas are usually restricted to the liver region but can be several

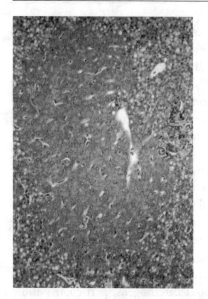
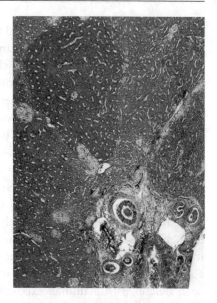

Figure 1.1 (left). Phenotypically altered focus of preneoplastic hepatocytes.
Figure 1.2 (right). Nodular proliferation of altered hepatocytes termed hepatocellular adenomas.

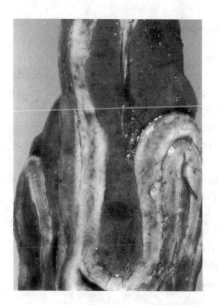
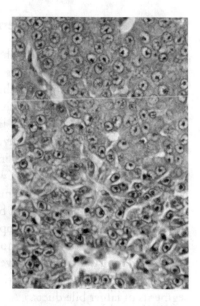

Figure 1.3 (left). Macroscopic appearance of multiple hepatocellular foci and
nodules in liver of a mature white sucker.
Figure 1.4 (right). Aggressive hepatocellular carcinoma on top, adjacent to
normal hepatic parenchyma.

cm in diameter. The morphological features of these various lesions correspond to similar stages of hepatocellular carcinogenesis observed in fish[24-26] and laboratory rodents[18,19] exposed to specific chemical hepatocarcinogens, and in humans with aflatoxin associated liver neoplasms.[28,29] However, in mammals, a similar series of stages occur in liver cancer development associated with viral, nutritional, or genetic causes. Unlike rodents exposed experimentally to genotoxic carcinogens,[18,19] white suckers with hepatic neoplasms have few preneoplastic focal proliferative lesions.[3,6,22]

Bile duct neoplasms are focal proliferations of epithelial cells with a ductular pattern of growth, and are classified as cholangiomas (Figure 1.5) or cholangiocarcinomas (Figure 1.6). These lesions are also termed bile duct adenomas and carcinomas, respectively. Focal atypical proliferations of bile duct epithelium that have not yet become expansive are not easily distinguished from hyperplastic epithelium, and cannot be enumerated in routine hematoxylin and eosin-stained tissue sections. Similar bile duct adenomas and carcinomas have been described in mammals exposed to various carcinogens, particularly those that lead to hyperplasia of epithelium of small (cholangiolar) or larger (cholangiar) ducts.[30] However, the chemically generated lesions are not morphologically distinguishable from bile duct neoplasms of other causes.

Inflammatory and Proliferative Responses

White suckers frequently exhibit hepatic inflammatory and proliferative lesions centered on the portal or ductular regions of the liver. These include cholangiohepatitis (Figure 1.7), cholangiar proliferation (hyperplasia) (Figure 1.8) or cholangiolar proliferation with interstitial fibrosis (cholangiofibrosis) (Figure 1.9). These changes can coexist in the same individual and are variably severe among different regions of the liver. These reactions appear to represent a spectrum of responses to focal duct injury, so, collectively, these conditions are referred to as chronic proliferative cholangiohepatitis. Cholangiohepatitis is characterized by a mixed inflammatory infiltrate centered on major and smaller portal tracts, particularly around bile ducts with proliferative changes. These infiltrates are lymphocyte and macrophage rich, suggestive of some antigenic stimulus within the bile duct system. Cholangiar proliferation is characterized by focal necrogenic injury and regenerative proliferation of epithelial cells of larger ducts. Not all ducts are affected, suggesting that the insult responsible causes necrosis in segments of larger bile ducts.

Cholangiofibrosis (Figure 1.9) is characterized by focal atrophy of lobar segments of liver, in which hepatocytes have been lost, while cholangiolar epithelium, interstitial fibroblasts, and macrophage centers enlarge. Frequently, these segments are at the extremities of elongated liver lobes and are grossly and histo-

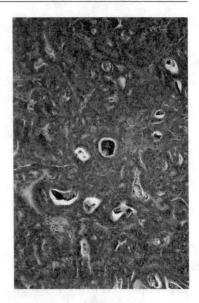

Figure 1.5 (left). Discrete cholangioma (bile duct adenoma).
Figure 1.6 (right). Aggressive cholangiocarcinoma (bile duct carcinoma).

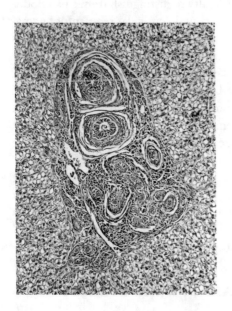
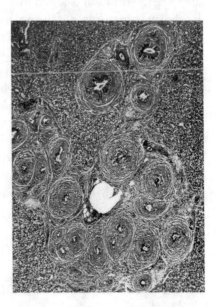

Figure 1.7 (left). Severe periportal leukocyte infiltration (cholangiohepatitis).
Figure 1.8 (right). Severe cholangiohepatitis with duct epithelial hyperplasia.

logically bile stained.[6,20] This pattern of response is characteristic of segmental bile duct obstructions. Since these obstructed lobes are usually found in livers with chronic proliferative cholangiohepatitis (above), they are considered a complication of primary injury to the ducts. However, the histological appearance needs to be distinguished from cholangiolar neoplasia by the presence of residual lobular architecture, macrophage centers, and bile retention, none of which are features of discrete neoplasms. Occasionally, neoplasms do occur within obstructed lobes. Focal histiocytosis is characterized by increased histological prominence of nodular macrophage centers, which increase in number, size, and pigmentation. Enlarged nodules of macrophages have prominent vacuolation and/or accumulation of yellow-brown to green-brown pigment. The latter pigmentation is typical in lobes with cholestasis, whereas the former is typical of phospholipid- and hemosiderin-rich lysosomal deposits that increase with age and hepatocyte turnover. Accordingly, these reactions are likely a consequence of chronic hepatic injury by insults responsible for chronic proliferative cholangiohepatitis and lobar atrophy. However, macrophage centers are more prominent in many old fish, some of which lack obvious inflammatory and proliferative reactions associated with the duct systems. This suggests that these focal histiocytic responses can increase during aging.

A proportion of white suckers in all sites have small discrete granulomas (Figure 1.10), sometimes in large numbers within the liver. Grossly, granulomas are spherical firm and pale foci that are difficulty to distinguish from small nodular cholangiomas. Some of these granulomas contain cuticular remnants of nematode larvae, but most contain amorphous, caseous necrotic debris. The granulomas are mostly delimited by a thin histiocytic encapsulation, without a strong leukocytic infiltrative response. A few fish in these sites have parasitic nematode and/or trematode larvae migrating through the liver. Nematodes are Spirurida larvae (Figure 1.11) consistent with *Acuariae* sp. found within necrotic tracts or nascent granulomas in the liver (Figure 1.9). Trematodes are larval and immature adult digenetic flukes (*Sanguinicola* sp.) (Figure 1.12) found within hepatic veins or free within necrotic tracts or granulomas in the hepatic parenchyma. In addition, some fish had large multicellular protozoan (*Myxosporea*) within the lumen of major bile ducts. Migratory nematodes and trematodes are usually associated with a local inflammatory response which proceeds to granulomas, whereas Myxosporea are consistently present within larger bile ducts that are not necessarily associated with inflammatory or proliferative responses.

EPIDEMIOLOGY OF LIVER DISEASE IN WHITE SUCKERS

Periodically since 1984, population samples of mature white suckers have been collected from various Ontario locations on the Great Lakes. These surveys,

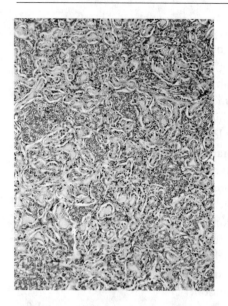
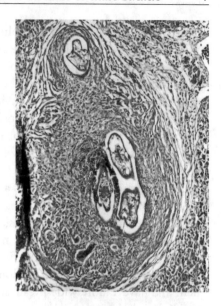

Figure 1.9 (left). Chronic cholangiolar hyperplasia and interstitial fibroplasia subsequent to segmental bile duct obstruction (cholangiofibrosis).
Figure 1.10 (right). Focal granuloma associated with cuticular nematode remnants.

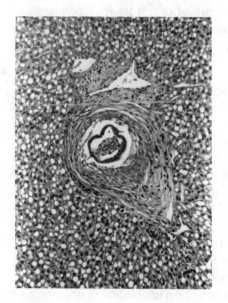
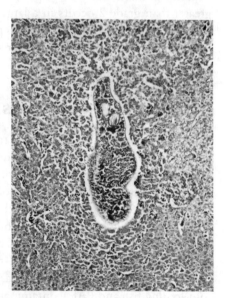

Figure 1.11 (left). Larval nematode in hepatic parenchyma consistent with *Acuariae* sp.
Figure 1.12 (right). Immature digenetic flukes (*Sanguinicola* sp.) in hepatic parenchyma.

conducted by Victor Cairns and John Fitzsimons of the Canadian Department of Fisheries and Oceans (DFO), are ongoing. Interim reports for some sites have been published[2,31] and others are pending the completion of repeated samples from the same sites. Some sites have been sampled by Ian Smith and David Rokosh of the Ontario Ministry of the Environment (MOE).[3,27] Livers from all these surveys have been subjected to detailed histopathological examination. Histopathological evaluation of all major liver lobes has been performed by M. Anthony Hayes for the Fish Pathology Laboratory, University of Guelph, and used to determine numbers of neoplastic lesions at various stages and frequency of fish affected. Hepatocellular neoplasms and prenoplastic altered foci occur in a low frequency (< 5%) in several locations in the more urban/industrialized regions of western Lake Ontario, but rarely in sites from eastern Lake Ontario, Lake Huron, and Lake Superior. Altered foci are more frequent than hepatomas and hepatocellular carcinomas, and some fish with these advanced neoplasms have preneoplastic altered foci. Cholangiomas and cholangiocarcinomas are more frequent and have a more widespread distribution.

Histopathological evaluation has also been used to assess severity and frequency of various nonneoplastic responses, each component rated on a 0–4 scale of increasing severity. These data indicate that chronic proliferative cholangiohepatitis characterized by component inflammatory and proliferative responses greater than grade 2 occurs in a large proportion of fish from all polluted and reference sites.[7,22] Numbers of various parasitic organisms and parasite-associated lesions in liver sections have also been determined for some of these sites (Figure 1.14).[7,22]

ETIOLOGICAL ASSOCIATIONS

Collectively, liver neoplasms are more frequent in western Lake Ontario locations in which habitat sediments are heavily contaminated with complex mixtures of polycyclic aromatic hydrocarbons (PAH) and many other contaminants of urban/industrial origin.[4,32-34] Bile of white suckers captured from these more contaminated locations contains much higher concentrations of metabolites with characteristic fluorescent properties of some groups of PAHs (Figure 1.13).[34,35] White suckers and brown bullheads exposed to PAH-contaminated sediments also have increased amounts of PAH-type DNA adducts.[4,36,37] PAHs are known carcinogens for fish,[4,24,25] and similar associations between liver neoplasms in bottom-feeding fish and PAH contamination have been established in other marine and freshwater habitats.[8,14-16,38,39] Collectively, these epidemiological associations are consistent with PAHs as risk factors for neoplasia in white suckers, but other constituents in these contaminated sediments[4,32-34] have not yet been excluded. In western Lake Ontario, liver neoplasms have been observed in white suckers

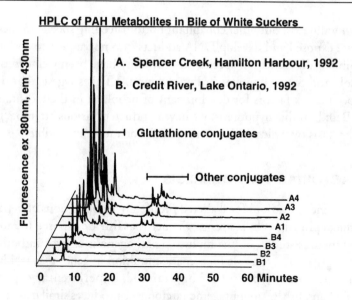

HPLC of PAH Metabolites in Bile of White Suckers

Figure 1.13. Comparison of fluorescent PAH-type metabolites[35] in bile of fish from contaminated site (Spencer Creek, Hamilton Harbour, western Lake Ontario) or from less contaminated Credit River, western Lake Ontario.

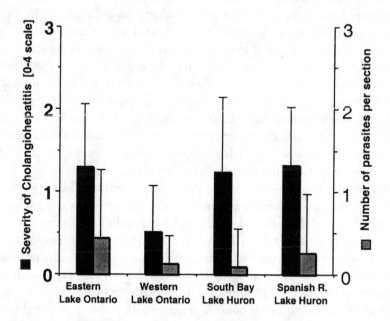

Figure 1.14. Comparison of severity of cholangiohepatitis (mean of 0–4 scale ± SD) and numbers of hepatic parasites per section (mean ± SD) in white suckers (N = 40–47 per site) between various sites in Lake Ontario and Lake Huron.

and brown bullheads, but other cohabitant bottom-feeding species in these habitats that are exposed and develop DNA adducts[37] do not appear to develop neoplasms. Similar differences in species' susceptibility have been well described for English sole and Starry flounder in Puget Sound.[13] This suggests that there are species-specific risk factors for development of neoplasia in these contaminated habitats. Based on this hypothesis, we have conducted various attempts to characterize the basis by which white suckers might be more susceptible.

MULTIFACTORIAL PATHOGENESIS

The presence of various focal altered proliferative populations of hepatocytes and bile duct epithelial cells is strong evidence that fish are exposed to mutagenic insults. Numerous studies on the multistep pathogenesis of liver and other neoplasms strongly implicate a role for somatic mutation in genes that regulate proliferation and phenotype.[18,19,40] For example, experimental exposure of mammals[18,19] and fish[24,25] to mutagenic carcinogens produces similar atypical prolific preneoplastic and neoplastic populations morphologically similar to those described in wild white suckers. However, specific mutated oncogenes implicated in mutagen-induced liver neoplasm of mammalian[19,40,41] and fish species[42] have not yet been characterized for white suckers. Accordingly, the hypothesis that such mutations are made by the lesions is based on correlative similarities.

White suckers are exposed to various PAHs in the more contaminated environments of western Lake Ontario,[32–37] and the sediments do contain various mutagenic constituents.[4,33] The hypothesis that some specific PAHs are responsible for somatic mutations in these neoplasms is based on circumstantial evidence, since PAHs have not been directly implicated. PAHs such as B[a]P are not strong, complete hepatocarcinogens, but can produce liver neoplasms in mammals and fish if there is an underlying cell proliferative influence.[43,44] Also, PAHs are not strong hepatocellular necrogens, so they are not among the more potent complete liver carcinogens which elicit postnecrotic hepatocellular regeneration that contributes the proliferative response required for initiation and promotion. However, during developmental or induced hepatocellular proliferation, some PAHs are carcinogenic in fish, as they are in mammals.[4,43] There remains a possibility that other contaminants including various persistent halogenated hydrocarbons might enhance carcinogenicity of PAHs by acting as postinitiating nongenotoxic promoters. For example, some PCBs and organochlorine pesticides promote liver cancer development in mammals[44] and fish[45] exposed to various PAHs, but in English sole, higher rates of exposure to PCBs does not increase susceptibility to liver carcinogenesis associated with PAHs.[13,39,46] Chronic cholangiohepatitis might also provide a necessary cocarcinogenic proliferative role. In mammals, inflammatory hepatitis caused by hepatitis B virus, in humans and other mammals, has been shown to be a potent risk factor for carcinogenicity of mutagens such as aflatoxin B1.[28,40,41] Also, hepatitis due to parasitic schistosomes in humans and rodents may also augment hepatocarcinogenicity of aflatoxin B1.[47,48] Accordingly, a plausible hypothesis is that liver neoplasms of white suckers are multifactorial; hepatitis could potentiate carcinogenicity of PAHs or other environmental pollutants to which the fish are exposed.

Inflammatory disease alone can be sufficient to elicit neoplastic responses in fish and mammals. The occurrence of cholangiomas in all contaminated and some reference habitats is perhaps in accordance with this hypothesis. Inflammatory cells are potent producers of endogenous mutagenic oxygen species which might be sufficient for mutation on a background proliferative response to hepatic injury underlying the chronic hepatic inflammatory disease.[49,50] However, this explanation does not easily apply to hepatocellular carcinogenesis in white suckers because altered foci and hepatocellular neoplasms are rare in various less contaminated habitats in which hepatitis is frequent. Hepatocytes are among the most proficient cells in neutralizing-reactive oxygen species because of their high concentrations of glutathione, glutathione transferases, glutathione peroxidases, and other antioxidants, so it seems less likely that inflammation-generated endogenous oxygen-derived mutagens could be responsible for the hepatocellular neoplasms. However, in mammals, activated phagocytes may also bioactivate xenobiotics, including PAHs.[51] Accordingly, environmental mutagen(s) restricted to the more contaminated sites are more plausible causes of mutations responsible for hepatocellular neoplasms, and inflammation could potentiate their mutagenicity.

While there are various reasons to implicate cholangiohepatitis in the pathogenesis of some liver neoplasms in white suckers, the causes of this condition are still unknown. Environmental mutagenic xenobiotics putatively responsible for initiating carcinogenesis in white suckers are unlikely causes of the chronic hepatitis because the latter is more widespread than are hepatocellular neoplasms.[7] Also, the severity of hepatitis does not correlate directly with the degree of urban-industrial contamination associated with the hepatocellular neoplasms.[7,20] This suggests that other factors endemic in all Great Lakes habitants are responsible for hepatitis. These factors could be endemic infectious pathogens or naturally occurring chemical hepatotoxins throughout all environments where inflammatory hepatic disease occurs. So far, we have identified various nematode, trematode, and protozoal parasites that elicit inflammatory reactions in the liver of white suckers but these have been found in a small proportion of fish with inflammatory and duct proliferative responses (Figure 1.14).[7,20] Implication of any of these parasites as causes of the more frequent chronic forms of hepatitis requires multiseason, multiyear, and multitissue epidemiological surveys in which infection rates for various stages of parasitic life cycles can be determined, but this has not been possible. Routinely, adult white suckers sampled for liver disease surveys have been captured upstream in various rivers during spring spawning migrations.[2,3] This approach provides a large population of sexually mature fish from selected geographic regions over a brief sampling period. However, it is expected that critical pathogenic events in the biology of these parasites occur within the lake habitats wherein white suckers feed on sediment organisms. Some white suckers have been captured occasionally from lake habitats by various netting procedures, and these fish do exhibit hepatitis and parasitic disease but numbers are insufficient for epidemiological determination of infection rates.

White suckers have substantial ability to conjugate and excrete PAHs, and when exposed, their bile contains many water-soluble PAH conjugates.[34,35] White suckers appear to have efficient biotransformation pathways for the hepatic detoxification and biliary excretion of PAHs such as benzo[a]pyrene (B[a]P). Experimentally administered ^{3}H-B[a]P is efficiently eliminated in the bile, and minute amounts of radioactivity are isolated as covalent adducts within hepatic protein and DNA.[6,7] Most B[a]P-like PAHs are bioactivated by hepatic cytochromes P-450 and conjugated by hepatic glutathione S-transferases (GST), glucuronyl transferases, or sulfotransferases in various mammals and fish.[35,39] In white suckers experimentally exposed to B[a]P, the predominant metabolites excreted in bile of white suckers are more polar than glucuronides and sulfates, and are refractory to beta glucuronidase and aryl sulfatase (Figure 1.13). These polar metabolites have HPLC elution properties analogous to glutathione conjugates of B[a]P.[6,7] Similar polar peaks with B[a]P-like fluorescence are abundant in bile of white suckers captured from more contaminated sites.[6,7,34] Also, purified cytosol prepared from liver of white suckers is efficient at conjugating B[a]P epoxides in a glutathione-dependent manner.[6] In comparison with various other fish species, white suckers are efficient at glutathione conjugation of B[a]P epoxides.[7] Genetic and acquired differences in cytochrome P-450-mediated activation of carcinogens can also explain diminished resistance in mammals.[29,52] Accordingly, it is expected that some diminution of resistance to PAHs may be required before PAHs can be plausibly implicated in the cause of these liver neoplasms.

We have evaluated the stability of the PAH detoxification phenotype in white suckers to address the possibility that it might be weakened in some ways. These investigations have shown that normal white sucker hepatocytes are strong expressers of various GST isoenzymes, some of which are likely involved in glutathione conjugation of PAH metabolites.[6,7,22,35] However, a minor proportion of preneoplastic hepatocytes, and a large majority of all hepatocellular and bile duct neoplasms are demonstrably GST-deficient.[22] This suggests that the native GST-dependent resistant phenotype of normal hepatocytes is reduced in those altered lesions that progress to later stages of neoplasia. This observation supports the hypothesis that PAHs and other GST-detoxified carcinogens could be responsible for some mutations required for neoplastic progression in those neoplastic populations in which GST-defenses are diminished.[22] In mice, repeated application of PAHs alone has been shown to increase the malignant conversion of benign papillomas to carcinomas.[53] PAHs might also be implicated in the initial mutations that putatively underlie the GST-deficient phenotypically altered hepatocellular foci, the likely progenitors of the GST-deficient later neoplasms. However, in rodent models, PAH-initiated preneoplastic and later populations usually have a GST-induced phenotype that is implicated in their selective proliferation under the influence of various toxin promoters.[18,54] The observations that promoted liver neoplasms either generated experimentally with PAHs in rainbow trout,[26] or present in wild-caught white suckers,[22] mostly have a deficient or normal rather than induced GST-phenotype suggest that they are not promoted by virtue of GST-dependent resistance, or that the GST-induced phenotype is not a common early manifestation of the initiated phenotype in fish.

In rodents in which liver neoplasms are promoted by various mitoinhibitory or mitogenic stimuli, the majority of preneoplastic and nodular lesions have a resistant phenotype characterized by elevations of various phase II detoxification enzymes, among which are some isozymes of GST.[18] In experimental hepatocarcinogenesis in rats, in which GST-induced resistant hepatocytes are generated with initiating and promoting carcinogenic chemicals, later lesions have selective reductions in mu class GST isozymes[22] that are most efficient in detoxifying reactive epoxide metabolites. Accordingly, there is a similarity between hepatocarcinogenesis in white suckers and rats in that those lesions that do progress to advanced stages of carcinogenesis have reductions in GSTs required for detoxification of some carcinogens. In addition, rat neoplasms have reduced glutathione peroxidase and superoxide dismutase.[22,55] These observations support the hypothesis that reduced expression of some phase II detoxification functions could contribute to an increased susceptibility of developing neoplasms to some agents involved in mutagenesis or progression.

We have considered the possibility that hepatitis might also impair detoxification of PAHs in white suckers. Immunoreactive hepatocellular GSTs are similar in livers with or without hepatitis,[6] but the possibility that specific GST isozymes might be selectively expressed has not been ruled out. However, several lines of evidence indicate that bile duct obstruction can alter the function of GSTs of white suckers. Lobes of liver with obvious obstruction and cholestasis have reduced GST activity.[6,7,20] Bile acids, especially chenodeoxycholic acid which accumulates in cholestatic liver is a competitive inhibitor of white sucker hepatic GST activity. Other substances such as hematin that accumulates in obstructed liver also inhibits GST activity in vitro.[6,20] While these observations indicate that GST function can be reduced by cholestasis, it is unknown if other inflammatory and proliferative processes responsible for obstruction can alter GST activity before obstruction occurs. ^{3}H-B[a]P administered to fish with predetermined lobar obstructions had less protein bound B[a]P than did unobstructed regions of liver, but amounts of B[a]P bound to DNA were similar.[6] These observations suggest that any alteration of GST activity associated with cholestasis does not potentiate the ability of B[a]P to form DNA adducts. However, those fish exposed to tracer doses of B[a]P had very low amounts of B[a]P bound to DNA. Accordingly, it is possible that the efficacy of B[a]P activation and conjugation is not impaired at exposure doses below a saturable threshold for GST function. Another possibility is that GST function might be more important in detoxification of other PAHs or mutagenic pollutants not examined. However, the occurrence of liver neoplasms, the end-product of putative carcinogenic environmental mutagens is not selective for regions of the liver most vulnerable to obstruction, and some occur in regions of the liver without histological evidence of previous duct obstruction. Collectively, these studies suggest that while obstructive liver disease can impair GST function, its role in facilitating carcinogenic initiating activity by improving GST-dependent detoxification is still uncertain, but evidently not necessary.

SIGNIFICANCE OF HEPATIC DISEASE IN WHITE SUCKERS

Epidemiological and mechanistic studies of neoplastic and nonneoplastic diseases in wild fish populations offer several opportunities for environmental quality assessment. They can indicate problem geographic regions and to some degree help with the assessment of severity of some impacts on unknown complex contaminant burdens. Until the specific causes are known, they may help to exclude or include classes of putative chemical contaminants among environmental risk factors of concern. Studies of these lesions has provided several insights into the mechanisms underlying multifactorial carcinogenesis. For example, they can help to determine which biological endpoints might be suitable as bioindicators of environmental impacts, either for regulatory management of problem contaminant discharges, or for regulating activities and resource utilization in various regions mapped according to tumor prevalence. Also, some endpoints might be useful in biomonitoring the status of impacted regions over time, particularly in evaluating the efficacy of prophylactic and remedial interventions for the improvement of environmental quality. Some of this knowledge might also be useful in measuring and mitigating risks to health and productivity of exposed aquatic populations, or health of terrestrial populations exposed to contaminants before and after they are deposited in the aquatic environment. Such applications based on fish disease data need to be made cautiously on the basis of critical interpretation of how these diseases occur and how they impact on health.

Identification of Problem Geographic Regions

It is clear that some classes of neoplastic liver disease occur more frequently in fish from some regions. For example, hepatocellular neoplasms are more frequent in white suckers and brown bullheads from western Lake Ontario.[1-7] Similar geographic associations between urban/industrial regions and liver neoplasms have been well documented for brown bullheads in Lake Erie,[5,8,9,11] for English sole in Puget Sound,[13,14] for winter flounder in Boston Harbour,[15] and for mummichog in Chesapeake Bay.[16] These strong correlations between various contaminants in upper sediments and the occurrence of hepatocellular neoplasms strongly suggest that the neoplasms are a direct reflection of the contaminated habitat, even if the specific etiological agents are not yet definitely known. Accordingly, the frequency of neoplasms in fish of similar age groups can thereby be used to map the extent and severity of the causative condition. However, there are some underlying weaknesses in this interpretation. For example, regional differences in tumor occurrence may reflect differences in cocarcinogenic, nutritional, or behavioral influences on uptake of or susceptibility to carcinogenic contaminants in sediments. If concurrent hepatitis influences susceptibility to PAHs, the prevalence of PAH-induced neoplasms would not necessarily correlate reliably with PAH exposure because not all fish are equivalently affected with hepatitis.

A further limitation in the use of tumor frequency data to map problem areas is the uncertainty concerning the range of movement of affected populations within the larger aquatic environment. It seems reasonable to expect that spawning populations captured for survey samples from particular rivers originate from nearby rather than remote regions in the lake, but the perimeters of the habitat regions in which sediment-borne exposure occurs are undefined. Accordingly, lower observed frequencies of neoplasms in relatively old fish captured from a particular river do not necessarily indicate that the environmental quality in the immediate vicinity is better than more distant sites in which frequencies are higher unless it can be established that the affected fish are not those that have ranged more widely. None-the-less, there are clearly some benefits to comparison of tumor prevalence among different regions. The main reliable uses of such data are in the primary recognition of larger regions in which problems occur. From this determination, additional investigations are required; for example, a comparison of equivalent lesions in fish captured directly from their deepwater habitat. Even when there are uncertainties concerning the range of movement, and contaminants to which they are exposed, it is possible to use these disease endpoints to determine the risk factors are having more or less impact in the same population over time. The observed decline in liver neoplasms in brown bullheads after reduced industrial discharges of PAHs is one example of this approach.[8,9] Such sequential studies are more readily conducted and interpreted for bullheads with high initial tumor prevalence and a lifespan of several years than for white suckers with much lower rates of neoplasms and a lifespan of up to 15–20 years. However, while hepatocellular neoplasms have so far been found only in white suckers from heavily contaminated sites, such lesions similar to those associated with PAH contamination in Lakes Erie and Ontario have been found frequently in brown bullheads in relatively pristine isolated lakes with low PAH contamination.[56,57]

Assessment of Severity of Impacts

An important application of health monitoring of aquatic species is the characterization of impacts that negatively influence productivity of animal species in a contaminated habitat. However, the neoplasms (Figure 1.15) and hepatitis (Figure 1.16) in white suckers in the Great Lakes do not have serious influences on the lifespan or age distribution of this species. These lesions are most severe in older fish, especially those greater than 10 years of age after several years of sexual maturity, so likely do not impair the ability of the populations to maintain themselves. White suckers with liver disease flourish in western Lake Ontario and at spawning, males and females have active ovarian and testicular development. However, there is mounting evidence that some still unidentified contaminants in bleached kraft pulp mill effluent have negative influences on reproduction in white suckers and other fish,[58] and some constituents of discharges are mutagenic in various in vitro systems (see Chapter 3).[59]

Identification of Specific Environmental Risk Factors

The pathogenesis of liver neoplasia observed in white suckers is similar to the processes observed in various species exposed to various carcinogens. Thus, few if any features of the neoplastic disease are expected to be useful in specific identification of causative environmental carcinogens. The complexities of multifactorial carcinogenesis in long-lived mammalian and fish species complicate the interpretation of epidemiological correlations. However, some aspects of the condition may be useful to rank the likelihood that various putative agents might be responsible. For example, the presence of hepatocellular preneoplastic and neoplastic lesions typical of those generated by mutagens during hepatocellular proliferation is evidence that fish are exposed to genotoxic insults. Also, the observation of GST-deficient phenotypes in later stage neoplasms suggests that some GST-detoxified carcinogens are involved in mutagenesis required for neoplastic progression. However, these considerations do not eliminate many candidate causes, in large part because little is known about susceptibility of white suckers to various xenobiotics.

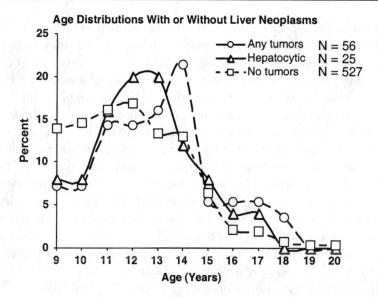

Figure 1.15. Age distribution of white suckers with liver neoplasms in comparison with those without neoplasms. Values are expressed as percentage ratios of fish collected from several sites in Lake Ontario.

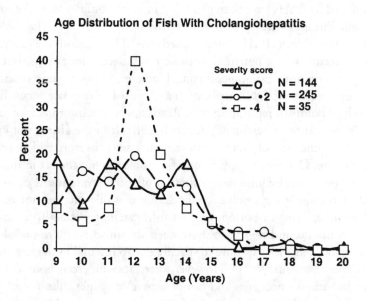

Figure 1.16. Age distribution of white suckers with cholangiohepatitis scores of 0 (unaffected), =2 (moderate to severe), or 4 (severe). Values are expressed as percentage ratios of fish collected from several sites in Lake Ontario.

More specific implication of a particular suspect requires evidence of exposure, uptake, and chemical effects relevant to carcinogenesis. Bottom-feeding fish such as white suckers, brown bullheads, and English sole consume many chemical contaminants in their food. Some of these are bioaccumulated in benthic organisms, while others are passive contaminants. Uptake can be readily demonstrated by various contaminants and their metabolites in tissues or fluids of exposed fish. For example, bottom-feeding fish exposed to PAHs in sediments excrete various PAH metabolites in bile,[34,35] and bioaccumulate various lipophilic halogenated hydrocarbons and heavy metal contaminants in their tissues. [32,33] Potent mutagenic chemicals tend to be reactive and more rapidly biotransformed and excreted, unlike the more stable persistent contaminants. Thus, the dosimetry of exposure is not necessarily a reliable basis for ranking chemicals as putative carcinogens in the absence of knowledge about the mechanisms of carcinogenesis. Various reactive xenobiotics are mutagenic carcinogens capable of initiating or increasing progression of liver neoplasms, whereas some less reactive persistent aromatic chemicals have tumor-promoting or cocarcinogenic effects.[18,53] It is possible that all of these influences might underlie the development of neoplasms of the types observed in bottom-feeding fish.

In recent years, some suspect environmental carcinogens have been more firmly implicated on the basis of evidence of promutagenic or mutagenic molecular lesions. For example, potentially promuatgenic DNA adducts typical of those produced by PAHs have been identified in brown bullheads, white suckers, walleye, and European carp in the Great Lakes,[4,36,37] and in English sole and winter flounder[39,60] from PAH-contaminated sites. These lesions are evidence of exposure to agents with a potential to cause carcinogenic mutations, but some exposed species do not develop neoplasms. However, promutagenic adducts are those not repaired before DNA replication and most DNA repair occurs during cell-cycle checkpoints in replicating cells. Accordingly, the interpretation of carcinogenic significance of persistent adducts is difficult in the absence of knowledge concerning the rates of DNA repair in replicating cells from which mutated populations arise. Definitive implication of a promutagenic adduct in the carcinogenic process would require demonstration of specific mutations of the type produced by the agent responsible for the adducts. While this has not yet been achieved for neoplasms in bottom-feeding fish, carcinogen-class specific point mutations in the ras and p53 genes have been shown to be more prevalent in neoplasms associated with exposure to specific environmental mutagens, including PAHs and aflatoxins.[19,40-42,61] However, there are numerous potential point mutations in various oncogenes and tumor suppressor genes. Absence of a particular mutated site does not eliminate a suspect carcinogen, and presence could result from other mutagens not under consideration.

Explanation of Mechanisms Underlying Multifactorial Carcinogenesis

The observation of clusters of particular neoplasms in geographically restricted habitats provides an opportunity to learn more about the interaction of various risk factors in the development of neoplasms. The original observation of liver neoplasms in rainbow trout was instrumental in determining that aflatoxins were liver carcinogens in other species, including humans.[24] Several potentially important concepts relating to multifactorial carcinogenesis have emerged from epidemiological studies of neoplasms in wild fish. For example, the demonstrated loss of GST phenotypes during progression of hepatocellular carcinomas in white suckers[22] parallels similar losses in resistance during carcinogenesis in rats[54] and humans.[47] The possibility that neoplastic progression involves prior development of a mutator phenoytpe in which constitutive resistance to mutations is diminished[62] has many implications for estimation of risks from exposure to potentially mutagenic environmental agents. Studies in white suckers suggest that loss of constitutive detoxification capacity either in preneoplastic populations[22] or in populations impacted by cholestasis[6] may contribute to an increased susceptibility to some carcinogens. Mutation susceptible phenotypes associated with loss-of-function in genes responsible for cell-cycle checkpoints, apoptosis, or DNA repair are increasingly being implicated in the pathogenesis of neoplasms in mammals.[62] These aspects of chemical carcinogenesis suggest that it is difficult to estimate carcinogenic potential of an environmental chemical exposure if the normal resistant phenotype is labile, as a consequence of germ-line or somatic mutations, or other adaptive influences associated with disease, diet, or other factors. More importantly, it suggests that carcinogenic or mutagenic potency of a chemical responsible for somatic mutations during neoplastic progression cannot be reliably assessed by effects produced in the constitutive mutation-resistant phenotype.

Design of Epidemiological Surveys of Neoplastic Diseases of White Suckers

The various surveys and mechanistic studies of neoplasms and other diseases of white suckers conducted over the past 15 years provide a basis for design of epidemiological approaches suitable for monitoring of impacts of contaminants to which they are still exposed. Spawning fish populations are the most readily accessible in early spring from various rivers, but this approach has logistical difficulties in that few sites can be examined in any one season. Also, the timing of collection is important because the population captured may vary in age distribution from day to day during the spawning runs. The samples collected from rivers should be regarded as representative of the full habitat range within the lake, rather than of the immediate vicinity of the site of capture, or the specific river drainage region. Evaluation of specific sites of concern within a lake requires additional samples of fish captured there at various times of the year, but this is an expensive endeavor and also subject to uncertainties about range of movement and population represented by the sample.

Because the prevalence of liver neoplasms in white suckers is low, and lesions occur in older fish (>8 years), the sample number should be sufficiently large to provide age-adjusted frequencies of various categories of lesions. The age and weight/age ratios are important dependent variables in these studies. For spawning fish, adjusted body weight should be used, determined by subtraction of gonad weight. Thus, a sample of approximately 200 large fish should provide age subsets (e.g., 100 fish >10 years of age) in the range in which lesions are readily found. Smaller numbers can be used in surveys of fish species with shorter lifespan or a high frequency of lesions.[63] In most situations, males and females are similarly affected, so sexes can be combined, but it is sometimes preferable to collect males only because the liver basophilia and hyperactivity of hepatocytes in spawning females makes the identification of preneoplastic hepatocyte lesions difficult in routine hematoxylin and eosin stained sections. If surveys are to be used for periodic evaluation of impacts in specific locations, the river-spawning populations should be compared at time intervals of at least 5–8 years, on the basis that the conditions observed represent lifetime experience in the habitat of concern. It is important to base these approaches on the theory that neoplasms might not occur in some species exposed to carcinogen contaminated sites,[4,64] or that some neoplasms might still occur after contaminant burdens are substantially reduced.

Most lesions of interest cannot be recognized by macroscopic examination of the liver, and the frequent inflammatory and cholestatic lesions are not evenly distributed among liver lobes. Accordingly, the tissue sampling procedure should be standardized and sufficient. Ideally, the entire liver, except for that required fresh for other determinations, should be collected and preserved in buffered formalin. Portions to be processed for microscopy should be obtained consistently from identical regions of each of the major lobes, so it is important that the necropsies are conducted carefully by those familiar with the normal structure and appearance of the liver. Metastatic liver neoplasms are rare in white suckers and other fish[21] so it is usually not necessary to collect other tissues. The identification and classification of various lesions is critical. Preneoplastic and neoplastic lesions must be differentiated from nonneoplastic lesions, but there are a number of interpretation problems. Discrete phenotypically altered preneoplastic hepatocyte populations must be differentiated from irregularities in storage of fat and glycogen in normal liver parenchyma; this is easier in livers from male fish. Bile duct epithelial proliferation associated with cholangitis, bile duct obstruction, and cholangiofibrosis must be differentiated from discrete cholangiomas and cholangiocarcinomas. These distinctions are best made on the basis that preneoplastic and neoplastic populations are clonal, and initially enlarge by expansive growth, during which the resident hepatic structures are displaced to the periphery, and at the perimeter, the clonal populations are sharply distinct from normal and hyperplastic cells. Because bile duct neoplasms can occur in association with cholangiohepatitis in fish from reference sites, it is necessary to differentiate neoplasms of hepatocellular origin from those of duct epithelial origin, and the severity of cholangiohepatitis and cholangiofibrosis should be evaluated on a quantitative basis in all fish. For most lesions, this distinction is readily made based on similarity of tumor cells with those of the normal population, but some advanced hepatocellular carcinomas become less differentiated or redifferentiated into ductal or tubular patterns that resemble those seen in cholangiocarcinomas.

Comparisons among fish population samples can be done in various ways. For neoplasms, frequencies of fish within particular age subsets with specific preneoplastic or neoplastic lesions are most easily determined. The preferred initial approach is to include fish according to the earliest observed lesion; for example, the percentage of all fish with preneoplastic and later lesions, or all fish with benign and later hepatocellular neoplasms can be used. However, fish often have multiple lesions, and the number of these lesions can reflect the magnitude of the carcinogenic influences. Thus, it is also useful to determine the total numbers of lesions of various categories, expressed and compared as mean number of lesions per fish. These latter approaches may provide a more sensitive endpoint for comparing sites in which the incidence of lesions is low.

SUMMARY AND CONCLUSIONS

White suckers *(Catostomus commersoni)* are bottom-feeding fish that inhabit northern freshwater lakes of North America. In the more contaminated regions of western Lake Ontario, a small proportion of older fish have various preneoplastic and neoplastic liver lesions. These fish in the more urban/industrial regions are exposed to much higher amounts of polycyclic aromatic hydrocarbons (PAHs) from contaminated sediments, but various other exposed species do not develop liver neoplasms. White suckers have efficient constitutive glutathione S-transferase (GST)-dependent hepatic detoxification and biliary excretion of PAHs. Most white suckers from various contaminated and reference sites in the Great Lakes have an idiopathic chronic proliferative cholangiohepatitis that may be a cofactor in the pathogenesis of liver neoplasms. While hepatocellular neoplasia is restricted to white suckers from the more contaminated regions, a low incidence of bile duct neoplasms occurs in white suckers from reference sites and may be associated with the chronic hepatobiliary disease. Bile duct obstruction in white suckers with this condition impairs GST activity, while regenerative ductal and hepatocellular proliferation, or acquired phenotypic deficiencies of GST in preneoplastic lesions, are predicted to increase hepatic susceptibility to various mutagenic xenobiotics, including PAHs. Neither the cholangiohepatitis nor liver neoplasia has an observable impact on age distribution of sampled populations of white suckers, so it is unlikely that these conditions are indicators of adverse impacts on the health of these fish populations. These studies indicate that epidemiological surveys of hepatocellular neoplasms and preneoplastic lesions of white suckers can be used to assess impacts of exposure to sediment carcinogens in some geographic regions.

ACKNOWLEDGMENTS

We are grateful for the generous cooperation of Victor Cairns, John Fitzsimons, Ian Smith, David Rokosh, Philip Byrne, Hugh Ferguson, Cam Portt, Chris Metcalfe, and Salem Rao, who provided material and expertise relevant to these studies. We are also grateful to Margaret Stalker, Trudy Kocal, Bette-Anne Quinn, Sonya Gordon, Cathy Thorn, and Tania Crane for contributions to the laboratory studies of neoplasms of white suckers. Research support for this work came from the Natural Sciences and Engineering Research Council of Canada, the Medical Research Council of Canada, the Ontario Ministry of the Environment, the World Wildlife Fund, the Canadian Department of Fisheries and Oceans, and the Ontario Ministry for Agriculture, Food and Rural Affairs.

REFERENCES

1. Dawe, C.J., R. Sonstegard, M.F. Stanton, D.E. Woronecki, and R.T. Reppert. Intrahepatic biliary duct neoplasms in *Catostomus commersoni. Prog. Exptl. Tumor Res.,* 20, pp. 195–204, 1976.
2. Cairns, V.W. and J.D. Fitzsimons. The occurrence of epidermal papillomas and liver neoplasms in white suckers *(Catostomus commersoni)* from Lake Ontario. *Can. Tech. Rep. Fish. Aquat. Sci.,* 1607, p. 151, 1988.
3. Hayes, M.A., I.R. Smith, T.L. Crane, T.H. Rushmore, T.E. Kocal, and H.W. Ferguson. Pathogenesis of skin and liver neoplasms in white suckers *(Catostomus commersoni)* from industrially polluted sites in Lake Ontario. *Sci. Total Environ.,* 94, pp. 105–123, 1990.
4. Metcalfe, C.D., G.G. Balch, V.W. Cairns, J.D. Fitzsimons, and B.P. Dunn. Carcinogenic and genotoxic activity of extracts from contaminated sediments in western Lake Ontario. *Sci. Total Environ.,* 94, pp. 125–141, 1990.
5. Black, J.J. and P.C. Baumann. Carcinogens and cancers in freshwater fishes. *Environ. Health Perspect.,* 90, pp. 27–33, 1991.

6. Kirby, G.M., M.J. Stalker, S. Gordon, F.J. van Schooten, and M.A. Hayes. Influences of chronic cholangiohepatitis and cholestasis on hepatic metabolism of benzo[a]pyrene in white suckers *(Catostomus commersoni)* from industrially-polluted areas of Lake Ontario. *Carcinogenesis,* 16, pp. 2923–2929, 1995.

7. Kirby, G.M. and M.A. Hayes. Significance of liver neoplasia in wild fish: Assessment of pathophysiologic responses of biomonitor species to multiple stress factors. *Can. Tech. Rep. Fish. Aquat. Sci.,* 1863, pp. 106–116, 1992.

8. Baumann, P.C., J.C. Harshbarger, and K.J. Hartman. Relations of liver tumors to age structure of brown bullhead populations from two Lake Erie tributaries. *Sci. Total Environ.,* 94, pp. 71–88, 1990.

9. Baumann, P.C. and J.C. Harshbarger. Decline in liver neoplasms in wild brown bullhead catfish after coking plant closes and environmental PAHs plummet. *Environ. Health Perspect.,* 103, pp. 168–170, 1995.

10. Stich, H.F. and A.B. Acton. The possible use of fish tumors in monitoring for carcinogens in the marine environment. *Prog. Exptl. Tumor Res.,* 20, pp. 44–54, 1976.

11. Couch, J.A. and J.C. Harshbarger. Effects of carcinogenic agents on aquatic animals: An environmental and experimental overview. *Environ. Carcinogenesis Revs.,* 3, pp. 63–105, 1985.

12. LeBlanc, G.A. and L.J. Bain. Chronic toxicity of environmental contaminants: Sentinels and biomarkers. *Environ. Health Perspect.,* 105, Suppl. 1, pp. 65–80, 1997.

13. Myers, M.S., C.M. Stehr, O.P. Olson, L.L. Johnson, B.B. McCain, S.L. Chan, and U. Varanasi. Relationships between toxicopathic hepatic lesions and exposure to chemical contaminants in English sole *(Pleuronectes vetulus),* starry flounder *(Platichthys stellatus),* and white croaker *(Genyonemus lineatus)* from selected marine sites on the Pacific Coast, USA. *Environ. Health Perspect.,* 102, pp. 200–215, 1994.

14. Myers, M.S., J.T. Landahl, M.M. Krahn, L.L. Johnson, and B.B. McCain. Overview of studies on liver carcinogenesis in English sole from Puget Sound; evidence for a xenobiotic chemical etiology. I: Pathology and epizootiology. *Sci. Total Environ.,* 94, pp. 33–50, 1990.

15. Murchelano, R.A. and R.E. Wolke. Neoplasms and nonneoplastic liver lesions in winter flounder, *Pseudopleuronectes americanus,* from Boston Harbor, Massachusetts. *Environ. Health Perspect.,* 90, pp. 117–126, 1991.

16. Vogelbein, W.K., J.W. Fournie, P.A. Van Veld, and R.J. Huggett. Hepatic neoplasms in the mummichog *Fundulus heteroclitus* from a creosote-contaminated site. *Cancer Res.,* 50, pp. 5978–5986, 1990.

17. Mikaelian, I., Y. de Lafontaine, C. Menard, P. Tellier, J. Harshbarger, and D. Martineau. Neoplastic and nonneoplastic hepatic changes in lake whitefish *(Coregonus clupeaformis)* from the St. Lawrence River, Quebec, Canada. *Environ. Health Perspect.,* 106, pp. 179–183, 1998.

18. Farber, E. The step-by-step development of epithelial cancer: from phenotype to genotype. *Adv. Cancer Res.,* 70, pp. 21–48, 1996.
19. Poole, T.M., T.A. Chiaverotti, R.A. Carabeo, and N.R. Drinkwater. Genetic analysis of multistage hepatocarcinogenesis. *Prog. Clin. Biol. Res.,* 395, pp. 33–45, 1996.
20. Kirby, G.M. Alterations in hepatic glutathione S-transferase-mediated resistance to environmental carcinogenesis in fish. PhD thesis, University of Guelph, 1991.
21. Hayes, M.A. and H.W. Ferguson. Neoplasms in Fish, in *Systemic Pathology of Fish.* H.W. Ferguson, Iowa State University Press. 1989, pp. 230–256.
22. Stalker, M.J., G.M. Kirby, T.E. Kocal, I.R. Smith, and M.A. Hayes. Loss of glutathione S-transferases in pollution-associated liver neoplasms in white suckers (*Catostomus commersoni*) from Lake Ontario. *Carcinogenesis,* 12, pp. 2221–2226, 1992.
23. Myers, M.S., L.D. Rhodes, and B.B. McCain. Pathologic anatomy and patterns of occurrence of hepatic neoplasms, putative preneoplastic lesions, and other idiopathic hepatic conditions in English sole (*Parophrys vetulus*) from Puget Sound, Washington. *J. Natl. Cancer Inst.,* 78, pp. 333–347, 1987.
24. Bailey, G.S., D.E. Goeger, and J.D. Hendricks. Factors Influencing Experimental Carcinogenesis in Laboratory Fish Models, in *Metabolism of Polycyclic Aromatic Hydrocarbons in the Aquatic Environment,* Varanasi, U., Ed., CRC Press, Boca Raton, FL, 1989, pp. 253–268.
25. Bunton, T.E. Experimental chemical carcinogenesis in fish. *Toxicol. Pathol.,* 24, pp. 603–618, 1996.
26. Kirby, G.M., M.J. Stalker, C.D. Metcalfe, T.E. Kocal, H.W. Ferguson, and M.A. Hayes. Expression of immunoreactive glutathione S-transferase in hepatic neoplasms induced by aflatoxin B1 or dimethylbenzanthracene in rainbow trout. *Carcinogenesis,* 11, pp. 2255–2257, 1990.
27. Smith, I.R., H.W. Ferguson, and M.A. Hayes. Epidermal papillomas epidemic in brown bullhead, *Ictalurus nebulosus* and white suckers *Catostomus commersoni* (Lecepede) populations from Ontario, Canada. *J. Fish Dis.,* 12, pp. 373–388, 1989.
28. Chen, C.J., M.W. Yu, and Y.F. Liaw. Epidemiological characteristics and risk factors of hepatocellular carcinoma. *J. Gastroenterol. Hepatol.,* 12, pp. S294–S308, 1997.
29. Kirby, G.M., G. Batist, L. Alpert, E. Lamoureux, R.G. Cameron, and M.A. Alaoui-Jamali. Overexpression of cytochrome P-450 isoforms involved in aflatoxin B1 bioactivation in human liver with cirrhosis and hepatitis. *Toxicol. Pathol.,* 24, pp. 458–467, 1996.
30. Sirica, A.E. Biliary proliferation and adaptation in furan-induced rat liver injury and carcinogenesis. *Toxicol. Pathol.,* 24, pp. 90–99, 1996.

31. Great Lakes Advisory Board. *1991 Report to the International Joint Commission.* International Joint Commission, United States and Canada, Windsor, Ontario, Canada, 1991, pp. 24–29.

32. Harlow, H.E. and P.V. Hodson. Chemical contamination of Hamilton Harbour: A review. *Can. Tech. Rep. Fish. Aquat. Sci.,* 1603, pp. 1–91, 1988.

33. Marvin, C.H., L. Allan, B.E. McCarry, and D.W. Bryant. Chemico/biological investigation of contaminated sediment from the Hamilton Harbour area of western Lake Ontario. *Environ. Mol. Mutagenesis,* 22, pp. 61–70, 1993.

34. Maccubbin, A.E., S. Chidambaram, and J.J. Black. Metabolites of aromatic hydrocarbons in the bile of brown bullheads *(Ictalurus nebulosus). J. Great Lakes Res.,* 14, pp. 101–108, 1988.

35. Kirby, G.M., J.R. Bend, I.R. Smith, and M.A. Hayes. The role of glutathione S-transferases in the hepatic metabolism of benzo[a]pyrene in white suckers *(Catostomus commersoni)* from polluted and reference sites in the Great Lakes. *Comp. Biochem. Physiol.,* 95C, pp. 25–30, 1990.

36. Dunn, B.P., J.J. Black, and A.E. Maccubbin. ^{32}P-postlabeling analysis of aromatic DNA adducts in fish from polluted areas. *Cancer Res.,* 47, pp. 6543–6548, 1987.

37. Maccubbin, A.E., J.J. Black, and B.P. Dunn. ^{32}P-postlabeling detection of DNA adducts in fish from chemically contaminated waterways. *Sci. Total Environ.,* 94, pp. 89–104, 1990.

38. Collier, T.K. and U. Varanasi. Hepatic activities of xenobiotic metabolizing enzymes and biliary levels of xenobiotics in English sole *(Parophrys vetulus)* exposed to environmental contaminants. *Arch. Environ. Contamin. Toxicol.,* 20, pp. 462–473, 1991.

39. Stein, J.E., W.L. Reichert, M. Nishimoto, and U. Varanasi. Overview of studies on liver carcinogenesis in English sole from Puget Sound; evidence for a xenobiotic chemical etiology. II: Biochemical studies. *Sci. Total Environ.,* 94, pp. 51–69, 1990.

40. Hollstein, M.C., C.P. Wild, F. Bleicher, S. Chutimataewin, C.C. Harris, P. Srivatanakul, and R. Montesano. p53 mutations and aflatoxin B1 exposure in hepatocellular carcinoma patients from Thailand. *Int. J. Cancer.* 53, pp. 51–55, 1993.

41. Lasky, T. and L. Magder. Hepatocellular carcinoma p53 G > T transversions at codon 249: The fingerprint of aflatoxin exposure? *Environ. Health Perspect.,* 105, pp. 392–397, 1997.

42. Fong, A.T., R.H. Dashwood, R. Cheng, C. Mathews, B. Ford, J.D. Hendricks, and G.S. Bailey. Carcinogenicity, metabolism and Ki-ras proto-oncogene activation by 7,12-dimethylbenz[a]anthracene in rainbow trout embryos. *Carcinogenesis,* 14, pp. 629–635, 1993.

43. Culp, S.J., D.W. Gaylor, W.G. Sheldon, L.S. Goldstein, and F.A. Beland. A comparison of the tumors induced by coal tar and benzo[a]pyrene in a 2-year bioassay. *Carcinogenesis,* 19, pp. 117–124, 1998.

44. Hayes, M.A. Carcinogenic and Mutagenic Effects of PCBs, in *Polychlorinated Biphenyls (PCBs): Mammalian and Environmental Toxicity.* Safe, S., Ed., Springer-Verlag, Hamburg, 1987, pp. 77–95.

45. Donohoe, R.M., Q. Zhang, L.K. Siddens, H.M. Carpenter, J.D. Hendricks, and L.R. Curtis. Modulation of 7,12-dimethylbenz[a]anthracene disposition and hepatocarcinogenesis by dieldrin and chlordecone in rainbow trout. *J. Toxicol. Environ. Health,* 54, pp. 227–242, 1998.

46. Myers, M.S., J.T. Landahl, M.M. Krahn, and B.B. McCain. Relationships between hepatic neoplasms and related lesions and exposure to toxic chemicals in marine fish from the U.S. West Coast. *Environ. Health Perspect.,* 99, pp. 7–15, 1991.

47. Kirby, G.M., C.R. Wolf, G.E. Nea, D.J. Judah, C.J. Henderson, P. Srivatanakul, and C.P. Wild. In vitro metabolism of aflatoxin B1 by normal and tumorous liver tissue from Thailand. *Carcinogenesis,* 14, pp. 2613–2620, 1993.

48. Kirby, G.M., P. Pelkonen, V. Vatanasapt, A.M. Camus, C.P. Wild, and M.A. Lang. Association of liver fluke *(Opisthorchis viverrini)* infestation with increased expression of cytochrome P450 and carcinogen metabolism in male hamster liver. *Mol. Carcinogenesis,* 11, pp. 81–89, 1994.

49. Floyd, R.A. Role of oxygen free radicals in carcinogenesis and brain ischemia. *FASEB J.,* 4, pp. 2587–97, 1990.

50. Wei, L., H. Wei, and K. Frenkel. Sensitivity to tumor promotion of SENCAR and C57BL/6J mice correlates with oxidative events and DNA damage. *Carcinogenesis,* 14, pp. 841–847, 1993.

51. Cavalieri, E.L. and E.G. Rogan. The approach to understanding aromatic hydrocarbon carcinogenesis: the central role of radical cations in metabolic activation. *Pharmacol. Ther.,* 55, pp. 183–199, 1992.

52. McGlynn, K.A., E.A. Rosvold, E.D. Lustbader, Y. Hu, M.L. Clapper, T. Zhou, C.P. Wild, X.L. Xia, A. Baffoe-Bonnie, D. Ofori-Adjei, et al. Susceptibility to hepatocellular carcinoma is associated with genetic variation in the enzymatic detoxification of aflatoxin B1. *Proc. Natl. Acad. Sci. U.S.A.,* 92, pp. 2384–2387, 1995.

53. Hennings, H., A.B. Glick, D.A. Greenhalgh, D.L. Morgan, J.E. Strickland, T. Tennenbaum, and S.H. Yuspa. Critical aspects of initiation, promotion, and progression in multistage epidermal carcinogenesis. *Proc. Soc. Exptl. Biol. Med.,* 202, pp. 1–8, 1993.

54. Stalker, M.J., T.E. Kocal, B.A. Quinn, S.G. Gordon, and M.A. Hayes. Reduced expression of glutathione S-transferase Yb2 during progression of chemically-induced hepatocellular carcinomas in Fischer 344 rats. *Hepatol.,* 20, pp. 149–158, 1994.

55. Stalker, M.J. Alterations in cytoprotective enzymes during progression of hepatic neoplasia. PhD thesis, University of Guelph, 1994.

56. Poulet, F.M., M.J. Wolfe, and J.M. Spitsbergen. Naturally occurring orocutaneous papillomas and carcinomas of brown bullheads *(Ictalurus nebulosus)* in New York State. *Vet. Pathol.,* 31, pp. 8–18, 1994.

57. Spitsbergen, J.M. and M.J. Wolfe. The riddle of hepatic neoplasia in brown bullheads from relatively unpolluted waters in New York State. *Toxicol. Pathol.,* 23, pp. 716–725, 1995.

58. Janz, D.M., M.E. McMaster, K.R. Munkittrick, and G.J. Van Der Kraak. Elevated ovarian follicular apoptosis and heat shock protein-70 expression in white sucker exposed to bleached kraft pulp mill effluent. *Toxicol. Appl. Pharmacol.,* 147, pp. 391–398, 1997.

59. Rao, S.S., B.A. Quinn, B.K. Burnison, M.A. Hayes, and C.D. Metcalfe. Assessment of genotoxic potential of pulp mill effluent using bacterial, fish and mammalian assays. *Chemosphere,* 31, pp. 3553–3566, 1995.

60. Varanasi, U., W.L. Reichert, and J.E. Stein. ^{32}P-postlabeling analysis of DNA adducts in liver of wild English sole *(Parophrys vetulus)* and winter flounder *(Pseudopleuronectes americanus). Cancer Res.,* 49, pp. 1171–1177, 1989.

61. Dragan, Y.P., J.R. Hully, J. Nakamura, M.J. Mass, J.A. Swenberg, H.C. Pitot. Biochemical events during initiation of rat hepatocarcinogenesis. *Carcinogenesis,* 15, pp. 1451–1458, 1994.

62. Loeb, L.A. Cancer cells exhibit a mutator phenotype. *Adv. Cancer Res.,* 72, pp. 25–56, 1998.

63. Baumann, P.C. Methodological considerations for conducting tumor surveys of fishes. *J. Aquat. Ecosyst. Health,* 1, pp. 127–133, 1992.

64. Baumann, P.C. The use of tumors in wild populations to assess ecosystem health. *J. Aquat. Ecosyst. Health,* 1, pp. 135–146, 1992.

Chapter 2

The Japanese Medaka (*Oryzias latipes*): An In Vivo Model for Assessing the Impacts of Aquatic Contaminants on the Reproductive Success of Fish

C.D. Metcalfe, M.A. Gray, and Y. Kiparissis

INTRODUCTION

Reproductive Success in Fish

Birge et al.[1] stated that "reproduction in aquatic animals usually is the most critical function affected by chronic toxicant stress." However, reproductive impairment in fish can occur as a result of exposure to chemical or physical stressors at all stages of the life cycle, including fertilization; embryonic development and hatching; sex differentiation; gametogenesis; final maturation; ovulation or spermiation; and spawning. Therefore, there are multiple endpoints of reproductive toxicity that could potentially be monitored in order to assess the impacts of aquatic contaminants on the reproductive success of fish (Table 2.1). There are several unique biological characteristics of the Japanese medaka (*Oryzias latipes*) that make it possible to monitor a large number of these parameters to assess reproductive health. We will review the endpoints that have been studied with Japanese medaka that could influence reproductive success, and we will critically evaluate the potential for using this fish species as an in vivo model for identifying compounds in the aquatic environment that impact upon the reproductive health of fish.

Endocrine Modulation

Recently, there has been a great deal of interest in the potential role of endocrine modulating substances in altering sexual development and reproductive

Table 2.1. Effects Induced by Chemical Contaminants in Fish Which Could Impact upon Reproductive Success.[a]

Endpoints at Early Life Stages of Fish	Endpoints in Mature Fish
Gametogenesis & gamete viability	Gonadal maturation
Fertilization rate	Changes in endocrine control
Egg characteristics after fertilization	Spermiation/ovulation
Hatching success*	Spawning/courtship*
Embryolarval development and toxicity*	Fecundity*
Sexual differentiation*	

[a] Endpoints which have been studied in in vivo studies with the Japanese medaka are indicated with an asterisk.

success in fish. Impacts observed in fish in the field that have been attributed to exposure to endocrine modulating substances include: (a) altered serum steroid levels and delayed gonadal maturation in lake whitefish (*Coregonus clupeaformis*) populations near pulp mills,[2] (b) masculinization of mosquitofish (*Gambusia affinis*) in streams downstream of pulp mills,[3] (c) synthesis of egg-yolk protein (i.e., vitellogenin) and reduced serum testosterone in male carp (*Cyprinus carpio*) captured near sewage treatment plants in rivers in the United States,[4] and (d) vitellogenin induction in male rainbow trout (*Oncorhynchus mykiss*) caged downstream of sewage treatment plants in rivers in Britain.[5] In many of these cases, the specific compounds causing these effects have not been identified. This is partly due to the lack of appropriate in vivo bioassay models for assessing the endocrine modulating potential of chemicals and complex mixtures to fish.[6]

Modulation of the endocrine processes controlled by the sex steroids may occur through several mechanisms. The current knowledge of endocrine control of sexual differentiation and reproduction in fish was summarized by Donaldson.[7] There is a complex system of positive and negative feedback mechanisms operating throughout the brain-pituitary-gonadal axis that controls the balance of sex steroids in both male and female fish (Figure 2.1). The balance between circulating levels of endogenous steroid hormones governs the sexual development and the reproductive cycles of fish and other vertebrates. It is important to realize however, that there are interspecific differences among teleosts in the activity and normal circulating levels of steroid hormones.[8] For instance, depending upon the fish species, either testosterone, 11-ketotestosterone, or 7α-hydroxytestosterone may be the primary endogenous male steroid.

Chemical contaminants may disrupt endocrine systems through a variety of mechanisms. For instance, the herbicide, atrazine, altered the estrous cycle in rats by increased circulating levels of endogenous estrogen through modulation of the hypothalamic-pituitary axis. However, the primary focus of endo-

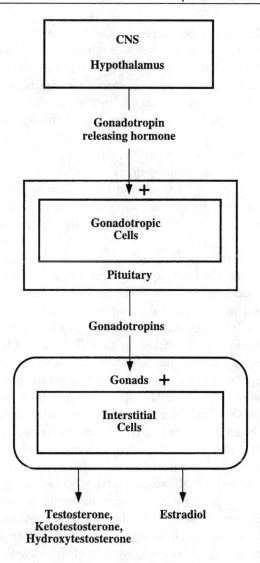

Figure 2.1. The brain-pituitary-gonadal axis of teleosts governing the release of sex steroid hormones by interstitial cells in the ovary or testis. A series of positive and negative feedback mechanisms control the levels of circulating plasma steroids.

crine modulation research has been on chemicals that act as "agonists," which mimic steroid hormones, and "antagonists," which inhibit the action of steroid hormones. Female sex steroid agonists are "estrogens" and male sex steroid agonists are "androgens."

Steroid agonists bind to hormone receptors in vertebrate cells and the resultant complex binds to specific responsive elements in the genome. Binding to these regions of the DNA causes a variety of biological responses that vary widely from organ to organ, and from species to species, depending upon the reproductive cycle of the organism. For instance in fish, ovarian secretion of the endogenous estrogen, 17β-estradiol results in binding of this ligand to the estrogen receptor (ER) in the hepatocytes of the liver and subsequent binding of this ligand-ER complex to responsive elements in the DNA (Figure 2.2), stimulating synthesis of the egg yolk protein, vitellogenin (VtG) which is incorporated into the developing eggs in the ovary. Exposure of fish to environmental estrogens has been detected through induction of VtG synthesis in male fish,[4,5] even though VtG synthesis is normally only observed in female fish that are affected by endogenous estrogens.[7]

Antagonistic activity occurs through either of two mechanisms of transcriptional inhibition: (a) Type I antagonism where a ligand binds to the hormone receptor but prevents binding to the DNA responsive element, and (b) Type II antagonism where a ligand binds to the receptor and the ligand-receptor complex binds to DNA, but fails to initiate transcription. For instance, several pesticides exhibit antiandrogenic activity, including the DDT metabolite, p,p'-DDE and the fungicide, vinclozolin.[9] It is possible that many of the adverse effects on reproductive development observed in fish that have been attributed to exposure to environmental estrogens may actually have occurred through antiandrogenic mechanisms.[9] At the moment, there are no in vivo bioassay techniques available for determining the antiandrogenic of environmental contaminants to fish.

Binding of 17β-estradiol to the estrogen receptor and testosterone to the androgen receptor in the gonads of fish and other vertebrates mediates sexual differentiation during the fetal and postnatal period of development.[7] Therefore, exposure of genotypic male fish to estrogens during early development may alter sex differentiation so that there is either complete sex-reversal to the female phenotype or development of an "intersex" condition.[10] Alternatively, exposure of fish to antiandrogens may lead to expression of the "default" female phenotype. Exposure to androgens may cause sex reversal of female fish to the male phenotype.[10]

The Japanese Medaka (*Oryzias latipes*)

The Japanese medaka is an oviparous freshwater killifish belonging to the Cyprinodont family. It is indigenous to southeastern Asia; common in rice paddies where it feeds extensively on mosquito larvae at the water surface. The wild strain of medaka is brownish-black in color, while cultivated varieties are light

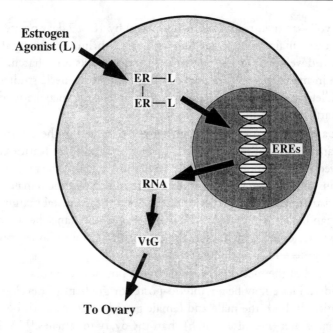

Figure 2.2. A model for induction of vitellogenin (VtG) synthesis in hepatocytes of telests involving binding of an estrogen agonist to a dimerized estrogen receptor (ER), binding of the agonist-ER complex with Estrogen Responsive Elements (EREs) in DNA, which stimulates transcription and synthesis of VtG.

gold, white, red, blue-black, or white variegated with black. Most of the early literature on medaka, dating back to the 1920s, can be attributed to the work of Japanese researchers with this species. Their popularity as a research organism is partly due to the ease with which they can be induced to breed. Medaka breed more freely than any other egg-laying aquarium fish, and a healthy female can produce more than 3,000 eggs in a single breeding season. Spawning is controlled by diurnal photoperiod and temperature, so breeding can be controlled in the lab with artificial light and manipulation of water temperature.[11]

The eggs are laid by the female after a very brief courtship, and spreading of milt by the male is concurrent with oviposition. Usually, 10–30 eggs are laid and are entangled by chorionic filaments near the female's vent. The cluster of eggs hangs from the female for several hours and these can be easily removed manually for subsequent studies. These eggs are valuable for embryological purposes because the egg chorion is clear, allowing development to be studied by examination of eggs under a dissecting microscope. At a temperature of 25°C, the time period between fertilization to hatch is 11–12 days and the fry completely absorb their yolk-sac (i.e., reach "swim-up" stage) by 18–19 days postfertilization. Adults are easily maintained by feeding with brine-shrimp and tropical fish food. Medaka can develop to sexual maturity within 6 weeks after hatch. Body pigmentation in medaka is a Mendelian trait of value for genetic studies. Since at least one allelic pair of genes for color is sex linked, the medaka lends itself to studies of artificially induced sex-reversals.[12,13]

There are subtle but clearly recognizable differences in the external secondary sex characteristics of male and female medaka, and these features are generally observed within 3 months of hatch. In mature medaka, the rays of the dorsal and anal fins of males are longer and thicker than those of the females, and have a characteristic notch at the posterior of the saw-toothed distal margin. The urinogenital papilla of the mature female is a paired protuberance between the anus and the oviduct opening, while on the male fish, it is a unilobed prominence between the anus and the urinogenital pore.

The gonad of the medaka is a single organ positioned medially beneath the swim bladder. There may be an anterio-posterior gradient of development and differentiation in both the male and female gonads, but the testis begins to develop in males later (i.e., after hatch) than the ovary in females.[13] In the female ovary, developing oocytes are surrounded by interstitial cells that promote ovarian maturation by releasing estradiol. Within the seminiferous tubules of the testis, spermatogonia develop into spermatocytes, which mature into spermatids and spermatozoa before spawning. The critical "window" for the differentiation of the gonad in male medaka occurs when fish are 6–8 mm in length,[13] which corresponds to approximately 13 days post-hatch. In contrast, sexual differentiation of the female gonad begins before hatching.[13] The medaka is characterized as a "differentiated gonochorist," while many other teleost species are "undifferentiated gonochorists." In differentiated gonochorists, the gonad develops directly into either a testis or an ovary, while in undifferentiated gonochorists, there is an indifferent gonad, which develops through an ovary-like stage, and then becomes either an ovary or a testis. The sex of a fish is more stable in differentiated gonochorists. Thus, no spontaneous intersex or sex reversal has ever been seen in medaka.[13]

The widespread use of medaka as a laboratory test species outside of Japan received impetus when it was determined that this species is particularly sensitive to the tumorigenic activity of chemical carcinogens.[14] Interest in this species as a toxicological model increased with the development of protocols for embryotoxicity testing with early life stages of medaka.[15] Most recently, this species shows promise as an in vivo model for studying the effects of endocrine modulating substances on the sexual development and reproductive performance of fish.

ENDPOINTS OF REPRODUCTIVE SUCCESS

Embryolarval Toxicity

Environmental chemicals often have the greatest toxicological impacts during the critical early stages of development and differentiation,[16] and fish mortalities at early life stages may have significant impacts on population recruitment. Embryolarval toxicity is an acute endpoint that involves exposing fish to chemicals at early life stages; beginning as early as the egg stage immediately after fertilization and generally extending to the period of development where fry begin exogenous feeding (i.e., "swim-up"). Alternately, the offspring of adult fish that have been exposed to chemicals could be assessed for early life-stage mortalities or developmental abnormalities; indicating "transgenerational" effects. Embryolarval toxicity tests with medaka extending from fertilization to swim-up can be completed over a period of 17–18 days at 25°C. The medaka embryolarval assay has been used to measure acute and developmental toxicities of several chemicals, including metals and organic compounds (Table 2.2).

Shi and Faustman[18] identified several reasons why the medaka is especially useful as a test organism for embryolarval toxicity:

- Under favorable conditions, sexually mature medaka can produce eggs daily.
- The embryo is highly visible, as the chorion of the medaka egg is transparent.
- The medaka is a small fish and therefore, many fish can be reared in a small area.
- Fertilized eggs hatch and develop relatively rapidly.
- Teratogenic effects in medaka embryos are comparable to those found in embryos of other fish species.
- Collection of eggs from adults is not invasive and does not involve sacrificing the fish.

For research specifically on the effects of endocrine-disrupting chemicals, the medaka embryolarval assay can be used as a prescreening tool to determine the lethal and sublethal toxicity levels of a test chemical, thereby avoiding situations where chronic toxicity studies end prematurely due to unexpected mortality. McKim[16] suggested that embryolarval endpoints with fish were robust enough to estimate maximum acceptable toxicant concentrations for chemicals without performing long-term toxicity tests involving partial or complete life-cycles.

The developmental stages of Japanese medaka embryos and larvae have been thoroughly described,[18,19] and a scoring system for developmental abnormalities has been developed specifically for medaka.[20] Toxicopathic lesions commonly

Table 2.2. Chemicals Which Induce Toxicity in Embryolarval Assays Using the Japanese Medaka.

Chemical	Reference
Industrial Chemicals	
2,3,7,8-TCDD and other dioxins and furans	15, 22, 23
Non-*ortho* PCB congeners (77, 81, 126)	22
Chlorinated diphenyl ethers	26
Nitroso-compounds (MNU, MNNG, DENA)	21,25
Alkylphenols (NP, OP)	17,27
Chlorophenols and nitrophenols	25,28
Hexachlorocyclohexane	24
Petroleum hydrocarbons	29
Toluene	30
Insecticides	
Carbamates (Carbaryl)	31
Organophosphates (Parathion, Malathion, Fenitrothion)	31,32
Organochlorines (DDT, Lindane, Aldrin)	31,33
Herbicides	
2,4-D	33
2,4,5-T	34
Metals	
Organic and inorganic mercury	35–38
Cadmium	39
Complex Environmental Mixtures	
Fish oil	40
Soil and sediment	41
Extract from Great Lakes fish	42

observed in embryolarval toxicity tests with medaka include slowed heart rate, hemostasis, hemorrhaging, tube hearts (i.e., undeveloped heart valves), pericardial edema, developmental arrest, and anisopthalmia (unequal eye diameters). Mortalities due to incomplete hatching are often observed, and after hatch, fry may experience reduced activity levels and may also fail to inflate swim bladders.[21,22]

Certain stages of medaka early life-stage development appear to be more sensitive to chemical insult due to either increased energy demands, or the particular mechanism of toxicity by the chemical. For instance, Wisk and Cooper[15,23] found stage-specific toxicity of 2,3,7,8-TCDD and other dioxins in medaka embryos at days 4 and 5 of development, coinciding with the period of liver formation. Mizell et al.[24] reported stage-dependent effects of hexachlorocyclohexane (HCH) on medaka embryos, and Gray et al.[17] found potential stage-specific toxicity at day 8 after fertilization in medaka embryos exposed to octylphenol.

There are several exposure protocols used in medaka embryolarval assays. Wisk and Cooper[23] and Harris et al.[22] utilized a static, nonrenewal, aqueous immersion protocol. The chemical is dissolved in a carrier solvent and added to a 2-mL glass vial. The solvent is evaporated, and then the chemical is redissolved in an aqueous rearing solution (i.e., saline solution with methylene blue to avoid fungal infection). Individual eggs are added to the vials within a few hours of fertilization, the vials are capped and incubated at 25°C for 17–18 days (i.e., to swim-up stage). Each egg or larva can be assessed for mortalities and developmental abnormalities in a matter of minutes. This method greatly facilitates daily observations without excessive disturbance to the embryos, and protects researchers from chemical exposure. Helmsetter and Alden[25] developed a method for topical application of a 0.1 μL volume of a chemical dissolved in DMSO to the surface of the medaka egg chorion. After 1 min exposure, the eggs are washed and placed in rearing solution. The advantage of this method is the ability to predict absolute toxicant doses using permeability factors, rather than relying on nominal exposure concentrations. Mizell et al.[24] developed a microinjection technique for medaka eggs which requires more sophisticated equipment and is more invasive than the methods described above.

It is also relatively simple to use these exposure protocols to assess the embryolarval toxicity of complex environmental mixtures. Cooper et al.[40] tested the embryolarval toxicity of fish oil that was suspected of being contaminated with dioxins and furans. The exposure protocol involved dilution of 1 μL of herring or cod liver oil in 1 mL of rearing solution, and individual medaka eggs were placed in 2-mL vials. The diluted fish oil caused stage-specific mortalities coinciding with development of the liver rudiment, as was previously seen with exposures to 2,3,7,8-TCDD.[23] Cooper et al.[41] also exposed medaka embryos to soils and sediments that were contaminated with dioxins and furans. Exposure to 1–10 mg of soil from Missouri and from Italy resulted in 100% mortality of medaka embryos. Harris et al.[42] tested extracts prepared from the tissues of Lake Ontario rainbow trout (*Oncorhynchus mykiss*) and were able to determine which contaminant fractions were responsible for medaka embryolarval toxicity. Fractionation of the extracts showed that most of the observed toxicity was due to non-*ortho* substituted PCBs and organochlorine compounds.

Embryos incubated at 25°C will generally hatch between 11 and 14 days after fertilization, with variations possible due to contaminant-related effects on the hatching process. Wisk and Cooper[23] observed that bioassays extending to 3 days posthatch are generally sufficient to establish whether medaka larvae will survive to swim-up. Also, swim bladder inflation generally occurs within 24 hours posthatch, so that if the embryo is "healthy," inflation should occur within 3 days posthatch.[21]

Vitellogenesis

Exposure to exogenous estrogens can induce vitellogenesis in male fish. Because vitellogenin is stimulated by binding of estrogen agonists to receptors in the hepatocytes of the fish liver, detectable levels of VtG in the plasma of male fish is a recognized biomarker for exposure to estrogenic compounds. For instance, plasma VtG concentrations increased by 100- to 1×10^6-fold in male rainbow trout exposed to alkylphenol compounds at μg/L levels for 3 weeks.[43] This induction of VtG was correlated with decreased testicular growth in the trout. Similar effects were observed in male trout deployed in cages downstream from discharges of sewage treatment plant effluents known to contain a variety of endocrine-disrupting compounds.[5,44]

Male fish have no natural depository for VtG, such as oocytes, so VtG tends to accumulate in the tissues. Herman and Kincaid[45] used estradiol treatments to reverse the sex of salmonids and observed high mortality among the male fish, likely due to excess VtG accumulation in the liver and kidney. VtG can also bind to calcium, resulting in potentially critical calcium deficiencies.[45] Adult male medaka exposed to high concentrations of estradiol developed swollen abdomens and experienced high mortality, which may have been associated with excess accumulation of VtG.[46] Wester and Canton[47] observed damage to the liver and kidney of male medaka associated with apparent vitellogenin production after exposure to the estrogen agonist, β-HCH.

Hamazaki and coworkers[46,48,49] conducted much of the fundamental research on the isolation and identification of VtG protein and other female-specific spawning proteins in medaka using immuno-histochemical techniques and electrophoretic analysis. Murata et al.[50] were able to show that all major constituents of the inner layer of the oocyte are formed in the liver of the medaka. There has been work done recently to develop a method for measuring VtG production in medaka. Because medaka are small fish and blood collection is difficult for measurement of plasma VtG levels, Nimrod and Benson[51] developed a method to measure VtG in hepatic cytosolic fractions. Exposure of adult male medaka to estradiol for 6 days resulted in a dose-dependent increase in cytosolic VtG levels in liver tissue.[51]

Sex Reversal and Intersex

Yamamoto and coworkers carried out numerous studies with medaka throughout the 1950s and 1960s to study the process of reversal of sexual differentiation.

A strain of medaka was developed with a gene coding for orange-red body color on the Y chromosome. This allows rapid identification of male fish within a few days of hatch, with 99% probablity.[52] This is especially important because the secondary sexual characteristics of fish are also under hormonal control,[53] and exposure to endogenous steroid hormones or endocrine disruptors can alter these phenotypic characters and lead to false identification of the genotypic sex of the fish. The use of the orange-red strain of medaka greatly facilitates experiments where sex reversals are induced since there is often ambiguity between the phenotypic indicators of sex.

Complete sex reversal in medaka has been observed after administration of both natural and synthetic steroid hormones. According to Yamamoto,[54] two conditions appear to be necessary for complete sex reversal. First, medaka must be exposed to a heterologous hormone (androgen for females, estrogen for males) during the critical stages of gonadal sex differentiation (i.e., just before hatch for females and just after hatch for males). Exposure after this period may induce temporary effects that may degenerate or disappear after exposure to the exogenous hormone ceases. Secondly, the dose of the heterologous hormone must be sufficient to induce complete sex reversal. Dosage levels below a threshold appear to induce an intersex condition. However, these assumptions must be studied more rigorously, especially for experimental systems where sex reversals or intersex conditions in medaka are induced with environmental estrogens or androgens.

Table 2.3 lists the endogenous hormones or chemicals that have induced complete sex reversal and/or an intersex condition in medaka. Note that sex reversals could not be adequately assessed in studies with environmental estrogens (NP, OP, β-HCH, o,p′-DDT, flavonoids), or with ethinylestradiol or testosterone as the strain of medaka used allowed only phenotypic identification and not genotypic verification of the sex of the fish. The intersex state observed in medaka, usually referred to as "testis-ova," represents a condition where both testicular and ovarian tissues occur simultaneously in the gonad (Figure 2.3). The data in Table 2.3 by a number of researchers indicate that several steroid estrogens and known environmental estrogens induce sex-reversals from male to female and/or testis-ova in male medaka. The small number of studies with steroid androgens indicate that these compounds induce sex-reversals from female to male and a similar intersex condition.

Yamamoto[12,13] was the first to note an anterio-posterior gradient of development within the intersex gonad of the medaka. Male fish exposed to exogenous estrogen agonists develop testis-ova in which testicular tissues tends to be localized to the anterior portion and ovarian tissues tend to be located in the posterior region of the gonad (Figure 2.1). It was suggested by Yamamoto and coworkers[55,56] that the anterior germ cells were more differentiated and under the influ-

Table 2.3. Chemicals Which Have Reversed the Sex and/or Induced an Intersex Condition in the Gonad in In Vivo Tests with Japanese Medaka.[a]

Chemical	Sex Reversal	Intersex	References
Endogenous Steroids			
Estrone	+	+	55*, 56+
17β-estradiol	+	+	13*, 57*, 58*, 59*
Hydroxyprogesterone	+	+	60*
Testosterone	n/a	+	61^
Synthetic Steroids			
Estriol	+	+	62*
Diethylstilbestrol	+	-	55*, 56+
Hexesterol, Euvestin	+	+	63*
Ethinylestradiol	n/a	+	61^
Methyltestosterone	+	+	12*
Environmental Estrogens			
β-HCH	n/a	+	47^
Nonylphenol	?[b]	+	27^
Octylphenol	n/a	+	64^
o,p′-DDT	n/a	+	65^
Flavonoids[c]	n/a	+	61^

[a] The vector for exposure to test compounds is identified by an asterisk (*) for dietary exposure, plus sign (+) for egg injection and a hat (^) for aqueous exposure.

[b] Change in sex ratio observed.

[c] Positive responses observed with chrysin, quercitin, naringenin, kaempferol, catechin, apigenin, flavone, flavonol, flavanone.

ence of endogenous hormones, whereas the posterior germ cells were still in an indifferent stage and susceptible to external influences. However, Yoshikawa and Oguri[10] also investigated the anterio-posterior gradient of differentiation in the medaka and found that the germ cells differentiated randomly within the gonad of both males and females, contrary to the hypothesis of Yamamoto.[55] Therefore, the mechanism for the observed gradients in testis-ova induction remains unclear at this time.

It appears that male medaka are more likely to develop testis-ova in response to exposure to estrogen agonists if exposure begins between 3 days and 1 week posthatch.[64] However, the formation of the gonad during early development of medaka and the role of primordial germ cells in this process has been studied by many researchers,[10,12,13,66,67] and there is no consensus on the optimal period for induction of testis-ova. Shibata and Hamaguchi[68] showed that there are germ cells in the testis of juvenile male medaka that retain their sexual bipotentiality

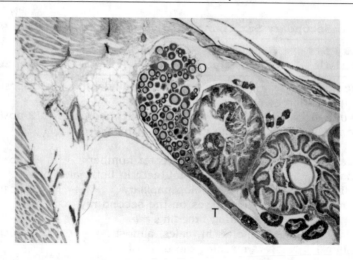

Figure 2.3. Testis-ova in a male medaka which was exposed from 1 day posthatch for 4 months to o,p'-DDT. Note the appearance of oocytes (O) in the anterior part of the gonad and testicular tissue (T) in the posterior part of the gonad. Magnification: X 400; H&E staining.

long after the gonad has differentiated into a testis. This observation is supported by studies where induction of testis-ova occurred in male medaka even when exposure to estrogen agonists began several weeks after gonadal differentiation.[47,59,64] It is interesting to note that external factors such as high temperatures that cause degeneration of the testes in adult male medaka promote the development of testis-ova that are induced by estradiol.[59]

Secondary Sex Characteristics

The medaka is a sexually dimorphic species, and mature male medaka are recognizable from the shape of the dorsal fin where a characteristic shallow notch exists in the posterior end formed by the separation of the hindmost ray from the other rays.[20,69] However, the most pronounced characteristic, the first to be identified in earlier studies, is the appearance of papillary processes in mature male medaka.[69] The rest of the dimorphic external morphological features such as body width, body color, papillary processes in the pectoral fin, size of the pelvic fins, size of the teeth and shape of the urogenital papillae (UGP) are more subtle. Some of the secondary sex characteristics become distinguishable when medaka reach 16–18 mm in length; others at 20 mm (dorsal and anal fins) or at 26 mm (papillary processes).[12] Other secondary sex characteristics such as nuptial coloration and papillary processes in the pectoral fin are temporary, appearing only

Table 2.4. Secondary Sex Characteristics in Japanese Medaka.[a,69–72]

Secondary Sex Characteristic	Description	Dependence
Body size and shape (permanent)	Males have greater body depth.	♂-positive
Nuptial coloration (temporary)	Greater number of leucophores in the caudal fin of males during breeding.	♂-positive
Teeth size (permanent)	Males have a greater number of large distal teeth in both jaws.	♂-positive
Pectoral fins (temporary)	Males develop papillary processes on the second ray during breeding.	♂-positive
Ventral fins (permanent)	Shorter in males, almost reaching the base of the anal fin.	♂-positive
Dorsal fin (permanent)	Larger in males with a distinguishable notch in the posterior end of distal edge. Rays are longer and thicker in males.	♂-positive
Anal fin (permanent)	Males have a longer parallelogram-like shaped fin with a notch in distal edge. Females have triangular shaped fin. In mature males, there are papillary processes on the posterior rays.	♂-positive
Urogenital pore (permanent)	More pronounced in mature females with pair of protuberances from the anus to the oviduct opening, giving a spherical shape. In males, a unilobed prominence between anus and urogenital pore.	♀-positive

[a] For each characteristic, there is information provided on the persistence (permanent or temporary) and hormonal dependence (male or female-positive).

during the breeding period, whereas the rest are permanent features that develop upon maturation (Table 2.4). The shapes of the anal and dorsal fins and the urogenital pore in male and female medaka are illustrated in Figures 2.4 and 2.5, respectively.

In the past, many experimental studies have been carried out to determine the effects of sex hormones on secondary sex characteristics.[69–72] Experiments in-

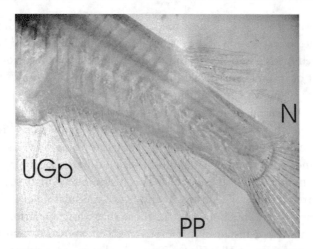

Figure 2.4. Some of the secondary sex characteristics of the male medaka, including the urogenital pore (UGp), anal fin with papillary processes (PP), and dorsal fin with notch (N).

Figure 2.5. Some of the secondary sex characteristics of the female medaka, including the urogenital pore (UGP), anal fin, and dorsal fin.

volving castration, ovariectomy, implantation and transplantation of gonads, and exposure to sex hormones indicated that in medaka, with the exception of the development of the UGP, which is a female-positive sex characteristic, the secondary sex features are male-positive since their development is controlled by the secretion of testicular hormones (Table 2.4). Administration of homologous and heterologous hormones to Japanese medaka resulted in either accelerated development or altered expression of the secondary sex characteristics.[13,62] For example, newly-hatched medaka exposed to androgens experienced a precocious arrhenoid development of the anal and dorsal fins at a length of 12–14 mm,[73] whereas the shape and size of the same fins in neuter and intersex medaka were in an intermediate status between the male and female phenotype.[55]

It is evident that the release of steroid hormones by the gonad regulates expression of the phenotypic sex under normal circumstances in the medaka, which is in accordance with the Jost paradigm of sexual differentiation in mammals.[74] Therefore, it might be hypothesized that alterations in secondary sex characteristics will occur as a result of exposure to endocrine-modulating chemicals at concentrations or dosages much lower than those that cause alterations to gonadal differentiation.[62,73] For instance, numerous medaka exposed to testosterone at a nominal concentration of 18 μg/L for 3 months developed papillary processes in the anal fin, regardless of their gonadal sex, whereas only one case of testis-ova was observed in these fish.[61] Also, papillary processes were observed in some female medaka exposed to the flavonoid compound, quercetin, but there was no occurrence of intersex in these treatments.[75]

Donaldson[7] mentioned that a delayed appearance of secondary sexual characteristics could be used as an index of reproductive impacts on fish exposed to pollutants. However, based on evidence of masculinized populations of mosquitofish (i.e., development of arrhenoid gonopodium in female fish) exposed to bleached kraft mill effluents[36] and to the phytosterols, β-sitosterol and stigmastanol,[76,77] it appears that alteration of the phenotypic sex in fish can also be used as an endpoint for assessing exposure to endocrine-modulating compounds. Although various endogenous and synthetic sex hormones have been used to manipulate the secondary sex characteristics in medaka, researchers have not taken advantage of this endpoint to establish endocrine-modulating effects of industrial effluents and xenobiotic compounds.

Full Life-Cycle and Transgenerational In Vivo Model

The medaka is an obvious choice for studying the effects of long-term exposure to endocrine-modulating compounds on reproductive success. The relatively rapid maturation of the medaka allows for full life-cycle studies within 3–4 months and transgenerational investigations within one year. Arcand-Hoy and Benson[78] recently proposed a model for investigating the effects of endocrine disruptors on medaka through first and second generations. This encompasses exposure of fish at embryolarval stages which are critical for gonadal differentiation, assessment of the reproductive potential of the fish as adults, and evaluation of survival and

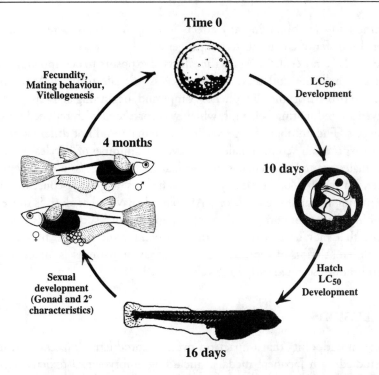

Figure 2.6. The life-cycle stages of the medaka and endpoints that can be used to assess the effects of chemicals on the reproductive success of fish.

developmental effects in the second generation. Figure 2.6 illustrates an experimental protocol for a full life-cycle test with medaka to assess the effects of chemicals on reproductive success. The biological assessments, including effects on mating behavior, fecundity, and survival, with corresponding biochemical and histological biomarkers are important endpoints for assessing the reproductive capacity of the fish.

There are four main components of this life-cycle model. Embryolarval toxicity tests can be used to determine the LC_{50} and the thresholds for sublethal developmental effects. The eggs from the embryolarval toxicity tests can also be exposed continuously through to maturity (i.e., 3–4 months), encompassing the entire life-cycle of the medaka. Exposure of medaka around the time of hatch is probably critical for assessing permanent responses to endocrine-modulating chemicals, as exposure at later life-stages may cause only temporary effects which regress after exposure ceases.[79] Other possible protocols involve: (a) exposing medaka to the test compound only during early life-stages, then exposing mature fish prior to spawning, (b) exposing medaka only when sexually mature to assess direct effects on reproduction, and (c) exposing female fish to chemicals prior to spawning and then observing any effects in offspring to assess responses to the maternal transfer of chemicals to the progeny.

Recently, Gray et al.[80] assessed the effects of exposure to octylphenol (OP) on the mating behavior and reproductive success of male medaka. Male medaka exposed to this estrogenic alkylphenol compound, beginning at 1 day posthatch, displayed altered mating behavior, which was correlated with reduced reproductive success. The number of "approaches" to females did not differ significantly between exposed and control males. However, the number of "circles" performed by the males around the females during mating did differ in a dose-dependent manner. The circling of females by males is the most aggressive courtship display in the mating ritual of the medaka.[80] Although behavioral analysis is more time-consuming than many other endpoints of reproductive success, this is a critical response that may be mediated by the neuroendocrine system. In vertebrates, sexual differentiation of reproductive and behavior patterns is affected by gonadal hormones secreted early in development.[81]

CONCLUSIONS

Several endpoints that impact on overall reproductive success in fish have been studied with Japanese medaka, including embryolarval toxicity, developmental abnormalities, differentiation of the gonad, development of secondary sexual characteristics, sex-reversals, vitellogenin induction in males, fecundity, and mating behavior. Several biological characteristics of this fish species make it ideal for studies on reproductive success. The medaka is relatively small, which aids in fish husbandry and in histological evaluation of all tissues in a single slide. The medaka spawns freely and lays a relatively large number of eggs. Once young are produced, a very short time period is required for the fish to reach sexual maturity. In addition, there is a great deal of information available on the basic life-history, genetics, histology, developmental biology, and mating behavior of the Japanese medaka.

However, there are some shortcomings to using medaka for studies on reproductive success and endocrine modulation. For instance, the small size of the fish makes analysis of serum steroids and serum VtG impossible. Also, the secondary sex characteristics of the species are relatively subtle in comparison to some other fish species, such as the fathead minnow. Therefore, identification of male and female medaka by untrained laboratory personnel can be difficult, especially if the fish are not fully sexually mature.

Most of the in vivo studies conducted with Japanese medaka have involved exposures to single test chemicals which have been either dissolved in water or administered through the diet. It remains to be seen whether test protocols with

Japanese medaka can be used to evaluate the reproductive effects of complex mixtures. Medaka have been used to assess the embryolarval toxicity of soils and sediments, herring oil, and fish extract, but to date, none of the other endpoints of reproductive success have been evaluated in studies with complex mixtures. It is possible that medaka could be used in in situ studies to evaluate the effects of aquatic contaminants on reproductive success. For instance, it should be possible to design test protocols with medaka to assess the effects of industrial and sewage treatment plant effluents on the reproductive development and success of fish.

REFERENCES

1. Birge, W.J., J.A. Black, and A.G. Westerman. Short-term fish and amphibian embryo-larval tests for determining the effects of toxicant stress on early life stages and estimating chronic values for single compounds and complex effluents. *Environ. Toxicol. Chem.,* 4, pp. 807–821, 1985.
2. Munkittrick, K.R., M.E. McMaster, C.B. Portt, G.J. Van Der Kraak, I.R. Smith, and D.G. Dixon. Changes in maturity, plasma sex steroid levels, hepatic mixed function oxygenase activity and the presence of external lesions in lake whitefish (*Coregonus clupeaformis*) exposed to bleached kraft mill effluent. *Can. J. Fish. Aquat. Sci.,* 49, pp. 1560–1569, 1992.
3. Howell, W.M., D.A. Black, and S.A. Bartone. Abnormal expression of secondary sex characters in a population of mosquitofish, *Gambusia affinis holbrooki*: evidence for environmentally induced masculinization. *Copeia* 1980, pp. 43–51, 1980.
4. Folmar, L.C., N.D. Denslow, V. Rao, M. Chow, D.A. Crain, J. Enblom, J. Marcino, and L.J. Guillette, Jr. Vitellogenin induction and reduced serum testosterone concentrations in feral male carp (*Cyprinus carpio*) captured near a major metropolitan sewage treatment plant. *Environ. Health Persp.,* 104, pp. 1096–1100, 1996.
5. Harries, J.E., D.A. Sheahan, S. Jobling, P. Matthiessen, P. Neall, E.J. Routledge, R. Rycroft, J.P. Sumpter, and T. Taylor. A survey of estrogenic activity in United Kingdom inland waters. *Environ. Toxicol. Chem.,* 15, pp. 1993–2002, 1996.
6. Hose, J.E. and L.J. Guillette. Defining the role of pollutants in the disruption of reproduction in wildlife. *Environ. Health Persp.,* (Suppl. 4) 103, pp. 87–89, 1995.
7. Donaldson, E.M. Reproductive indices as measures of the effects of environmental stressors in fish. *Amer. Fish. Soc. Symp.,* 8, pp. 109–122, 1990.
8. Campbell, P.M. and T.H. Hutchinson. Wildlife and endocrine disruptors: Requirements for hazard identification. *Environ. Toxicol. Chem.,* 17, pp. 127–135, 1998.

9. Monosson, E., W.R. Kelce, M. Mac, and L.E. Gray. Environmental Antiandrogens: Potential Effects on Fish Reproduction and Development, in *Chemically-induced Alterations in Functional Development and Reproduction of Fishes*, Rolland, R.M., M. Gilbertson, and R.E. Peterson, Eds., Proc. from a session at the 1995 Wingspread Conference, July 21–23, Racine, WI, SETAC Technical Publ. Series, 1997, pp. 53–60.

10. Yoshikawa, H. and M. Oguri. Gonadal sex differentiation in the medaka, *Oryzias latipes*, with special reference to the gradient of differentiation. *Bull. Jap. Soc. Sci. Fish.*, 45, pp. 1115–1121, 1979.

11. Kirchen, R.V. and W.R. West. Teleostean development. *Carolina Tips*, 32, pp. 33–45, 1969.

12. Yamamoto, T.-O. Artificially induced sex-reversal in genotypic males of the medaka (*Oryzias latipes*). *J. Exp. Zool.*, 123, pp. 571–578, 1953.

13. Yamamoto, T.-O. Artificial induction of functional sex-reversal in genotypic females of the medaka (*Oryzias latipes*). *J. Exp. Zool.*, 137, pp. 227–236, 1958.

14. Metcalfe, C.D. Tests for predicting carcinogenicity in fish. *CRC Crit. Rev. Aquat. Sci.*, 1, pp. 111–127, 1989.

15. Wisk, J.D. and K.R. Cooper. 1990. The stage specific toxicity of 2,3,7,8-tetrachlorodibenzo-*p*-dioxin in embryos of the Japanese medaka (*Oryzias latipes*). *Environ. Toxicol. Chem.*, 9, pp. 1159–1169.

16. McKim, J.M. Evaluation of tests with early life stages of fish for predicting long-term toxicity. *J. Fish. Res. Board Can.*, 34, pp. 1148–1154, 1977.

17. Gray, M.A., and C.D. Metcalfe. Toxicity of octylphenol to early life stages of Japanese medaka, *Oryzia latipes. Aquatic Toxicol.* (Accepted).

18. Shi, M. and E.M. Faustman. Development and characterization of a morphological scoring system for medaka (*Oryzias latipes*) embryo culture. *Aquat. Tox.*, 15, pp. 127–140, 1989.

19. Iwamatsu, T. Stages of normal development in the medaka *Oryzias latipes*. *Zoological Sci.*, 11, pp. 825–839, 1994.

20. Kirchen, R.V. and W.R. West. The Japanese Medaka: Its Care and Development. Carolina Biological Co., Burlington, NC, 1979.

21. Marty, G.D., J.M. Nunez, D.J. Lauren, and D.E. Hinton. Age-dependent changes in toxicity of *N*-nitroso compounds to Japanese medaka (*Oryzias latipes*) embryos. *Aquat. Tox.*, 17, pp. 45–62, 1990.

22. Harris, G.E., Y. Kiparissis, and C.D. Metcalfe. Assessment of the toxic potential of PCB congener 81 (3,4,4′,5-tetra-chlorobiphenyl) to fish in relation to other non-ortho substituted PCB congeners. *Environ. Toxicol. Chem.*, 13, pp. 1393–1404, 1994.

23. Wisk, J.D. and K.R. Cooper. Comparison of the toxicity of several polychlorinated dibenzo-p-dioxins and 2,3,7,8-tetrachloro-dibenzofuran in embryos of the Japanese medaka (*Oryzias latipes*). *Chemosphere*, 20, pp. 361–377, 1990.

24. Mizell, M., E. Roming, W. Hartley, and A. Thiyagarajah. Sex on the brain but the heart is not really in it: Developmental heart defects associated with aquatic pollution and microinjection of hexachloro-cyclohexane into the Japanese medaka embryo. *Biol. Bull.*, 189, pp. 196–197, 1995.

25. Helmsetter, M.F. and R.W. Alden, III. Passive trans-chorionic transport of toxicants in topically treated Japanese medaka (*Oryzias latipes*) eggs. *Aquat. Tox.*, 32, pp. 1–13, 1995.

26. Metcalfe, C.D., T.L. Metcalfe, J.A. Cormier, S.Y. Huestis, and A.J. Niimi. Early life stage mortalities in Japanese medaka (*Oryzias latipes*) exposed to polychlorinated diphenyl ethers. *Environ. Toxicol. Chem.*, 16, pp. 1749–1754, 1997.

27. Gray, M.A. and C.D. Metcalfe. Induction of testis-ova in Japanese medaka (*Oryzias latipes*) exposed to p-nonylphenol. *Environ. Toxicol. Chem.*, 16, pp. 1082–1086, 1997.

28. Waterman, A.J. Effects of 2,4-dinitrophenol on the early development of the teleost *Oryzias latipes*. *Biol. Bull.*, 78, pp. 29–36, 1940.

29. Leung, T.S. and R.V. Bulkley. Effects of petroleum hydrocarbons on length of incubation and hatching success in the Japanese medaka. *Bull. Environ. Toxicol.*, 23, pp. 236–245, 1979.

30. Stoss, F.W. and T.A. Haines. The effects of toluene on embryos and fry of the Japanese medaka, *Oryzias latipes*, with a proposal for rapid determination of maximum acceptable toxicant concentration. *Environ. Pollut.*, 139, pp. 13–22, 1979.

31. Solomon, H.M. and J.S. Weis. Abnormal circulatory development in medaka caused by the insecticides carbaryl, malathion, and parathion. *Teratology.* 19, pp. 51–60, 1979.

32. Takimoto, Y., S. Hagino, H. Yamada, and J. Miyamoto. The acute toxicity of fenitrothion to killifsh (*Oryzias latipes*) at twelve different stages of its life history. *J. Pest. Sci.*, 9, pp. 463–472, 1984.

33. Crawford, R.B. and A.M. Guarino. Effects of environmental toxicants on development of a teleost embryo. *J. Environ. Pathol. Toxicol. Oncol.*, 6, pp. 185–196, 1985.

34. Schreiweis, D.O. and G. Murray. Cardiovascular malformations in *Oryzias latipes* embryos treated with 2,4,5-trichlorophenoxyacetic acid (2,4,5-T). *Teratology* 14, pp. 287–295, 1976.

35. Dial, N.A. Some effects of methylmercury on development of the eye in medaka fish. *Growth*, 42, pp. 309–317, 1978.

36. Heisinger, J.F. and W. Green. Mercuric chloride uptake by eggs of the ricefish and resulting teratogenic effects. *Bull. Environ. Contam. Toxicol.*, 14, pp. 665–675, 1975.

37. Akiyama, A. Acute toxicity of two organic mercury compounds to the teleost, *Oryzias latipes*, in different stages of development. *Bull. Jpn. Soc. Sci. Fish.*, 36, pp. 563–572, 1970.

38. Sakaizumi, M. Effect of inorganic salts on mercury-compound toxicity to the embryos of the medaka, *Oryzias latipes. J. Fac. Sci. Univ. Tokyo,* 14, pp. 369–380, 1980.

39. Michibata, H. Effects of water hardness on the toxicity of cadmium to the egg of the teleost, *Oryzias latipes. Bull. Environ. Contam. Toxicol.,* 27, pp. 187–195, 1981.

40. Cooper, K.R., H. Liu, P.A. Bergqvist, and C. Rappe. Evaluation of Baltic herring and Icelandic cod liver oil for embryo toxicity, using the Japanese medaka (*Oryzias latipes*) embryo larval assay. *Environ. Toxicol. Chem.,* 10, pp. 707–714, 1991.

41. Cooper, K.R., J. Schell, T. Umbreit, and M. Gallo. Fish-embryo toxicity associated with exposure to soils and sediments contaminated with varying concentrations of dioxins and furans. *Mar. Environ. Res.,* 35, pp. 177–180, 1993.

42. Harris, G.E., T.L. Metcalfe, C.D. Metcalfe, and S.Y. Huestis. Toxicity of contaminants extracted from Lake Ontario rainbow trout to embryos of the Japanese medaka (*Oryzias latipes*). *Environ. Toxicol. Chem.,* 13, pp. 1405–1414, 1994.

43. Jobling, S., D. Sheahan, J.A. Osborne, P. Matthiessen, and J.P. Sumpter. Inhibition of testicular growth in rainbow trout (*Onchorhynchus mykiss*) exposed to estrogenic alkylphenolic chemicals. *Environ. Toxicol. Chem.,* 15, pp. 194–220, 1996.

44. Harries, J.E., D.A. Sheahan, S. Jobling, P. Matthiessen, P. Neall, J.P. Sumpter, T. Taylor, and N. Zaman. Estrogenic activity in five United Kingdom rivers detected by measurement of vitellogenesis on caged male trout. *Environ. Toxicol. Chem.,* 16, pp. 534–542, 1997.

45. Herman, R.L. and H.L. Kincaid. Pathological effects of orally administered estradiol to rainbow trout. *Aquaculture,* 72, pp. 165–172, 1988.

46. Hamazaki, T.S., I. Iuchi, and K. Yamagami. Production of a "spawning female-specific substance" in hepatic cells and its accumulation in the ascites of the estrogen-treated adult fish, *Oryzias latipes. J. Exp. Zool.* 242, pp. 325–332, 1987.

47. Wester, P.W. and J.H. Canton. Histopathological study of *Oryzias latipes* (medaka) after long-term β-hexachlorocyclohexane exposure. *Aquat. Tox.,* 9, pp. 21–45, 1986.

48. Hamazaki, T.S., I. Iuchi, and K. Yamagami. Purification and identification of vitellogenin and its immunohistochemical detection in growing oocytes of the teleost, *Oryzias latipes. J. Exp. Zool.,* 242, pp. 333–341, 1987.

49. Hamazaki, T.S., Y. Nagahama, I. Iuchi, and K. Yamagami. A glycoprotein from the liver constitutes the inner layer of the egg envelope (*zona pellucida interna*) of the fish, *Oryzias latipes. Dev. Biol.,* 133, pp. 101–110, 1989.

50. Murata K., T.S. Hamazaki, I. Iuchi, and K. Yamagami. Spawning female-specific egg envelope glycoprotein-like substances in *Oryzias latipes. Dev. Growth Different.,* 33, pp. 553–562, 1991.

51. Nimrod, A.C. and W.H. Benson. Assessment of Estrogenic Activity in Fish, in *Chemically-induced Alterations in Functional Development and Reproduction of*

Fishes, Rolland, R.M., M. Gilbertson, and R.E. Peterson, Eds., Proc. from a session at the 1995 Wingspread Conference, July 21–23, Racine, WI, SETAC Technical Publ. Series, 1997, pp. 87–100.

52. Yamamoto, T.-O. Hormonic factors affecting gonadal sex differentiation in fish. *Gen. Comp. Encocrinol.*, Suppl 1, pp. 341–345, 1962.

53. Yamamoto, T.-O. Sex Differentiation, in *Fish Physiology, Vol. III*. Hoar, W.S. and D.J. Randall, Eds., Academic Press, New York, 1969, pp. 117–175.

54. Yamamoto, T.-O. *Medaka (Killifish): Biology and Strains.* Keigaku Publishing Co., Yugaku-sha, Tokyo, Japan, 1975.

55. Yamamoto, T.-O. The effects of estrone dosage level upon the percentage of sex reversals in genetic male (XY) of the medaka (*Oryzias latipes*). *J. Exptl. Zool.* 141, pp. 133–154, 1959.

56. Hishida, T. Reversal of sex differentiation in genetic male of the medaka (*Oryzias latipes*) by injecting estrone-16-C^{14} and diethylstilbestrol (monoethyl-1-C^{14}) into the egg. *Embryologia*, 8, pp. 234–246, 1964.

57. Yamamoto, T.-O and N. Matsuda. Effects of estradiol, stilbestrol and some alkyl-carbonyl androstanes upon sex differentiation in the medaka, *Oryzias latipes. Gen. Comp. Endocrinol.*, 3, pp. 101–110, 1963.

58. Egami, N. Production of testis-ova in adult males of *Oryzias latipes*, I. Testis-ova in fish receiving estrogens. *Jap. J. Zool.*, 11, pp. 353–365, 1955.

59. Egami, N. Production of testis-ova in adult males of *Oryzias latipes*. VI. Effect on testis-ovum production of exposure to high temperature. *Annotat. Zool. Japonenses*, 29, pp. 11–18, 1956.

60. Yamamoto, T.-O. Effects of 17α-hydroxyprogesterone and androstenediol upon sex differentiation in the medaka, *Oryzias latipes. Gen. Comp. Endocrinol.*, 10, pp. 8–13, 1963.

61. Metcalfe, C.D., M.A. Gray, and T.L. Metcalfe. 1997. Induction of testis-ova in medaka by exposure to estrogenic compounds. *Proc. of 18th Annual Meeting of Society of Environmental Toxicology and Chemistry*, Nov. 16–20, 1997, San Fransisco, CA, p. 137.

62. Yamamoto, T.-O. Estriol-induced XY females of the medaka (*Oryzias latipes*) and their progenies. *Gen. Comp. Endocrin.*, 5, pp. 527–533, 1965.

63. Yamamoto, T.-O. Unpublished; cited in Yamamoto, Reference No. 53.

64. Gray, M.A., A.J. Niimi, and C.D. Metcalfe. Factors affecting the development of testis-ova in medaka, *Oryzias latipes*, exposed to octylphenol. *Environ. Toxicol. Chem.* (Accepted).

65. Metcalfe, T.L., C.D. Metcalfe, and A.J. Niimi. Effects of o,p-DDT on survival and gonadal development of Japanese medaka, *oryzias latipes:* Direct exposure and a transgenerational study. *Environ. Toxicol. Chem.* (Submitted).

66. Gamo, H. On the origin of germ cells and formation of gonad primordia in the medaka, *Oryzias latipes. Jap. J. Zool.*, 13, pp. 101–115, 1961.

67. Satoh, N. and N. Egami. Sex differentiation of germ cells in the teleost, *Oryzias latipes*, during normal embryonic development. *J. Embryol. Exptl. Morph.*, 28, pp. 385–395, 1972.

68. Shibata, N. and S. Hamaguchi. Evidence for the sexual bipotentiality of spermatogonia in the fish, *Oryzias latipes. J. Exptl. Zool.*, 245, pp. 71–77, 1988.

69. Egami, N. Secondary Sexual Characters, in *Medaka (Killifish): Biology and Strains*, Yamamoto, T., Ed., Keigaku Publishing Co., Yugaku-sha, Tokyo, Japan, 1975, pp. 109–125.

70. Egami, N. Note on sexual difference in the shape of the body in the fish, *Oryzias latipes. Annot. Zool. Jap.*, 32, pp. 59–64, 1959.

71. Egami, N. Notes on sexual difference in size of teeth of the fish, *Oryzias latipes. Jap. J. Zool.*, 12, pp. 65–69, 1956.

72. Egami, N. Occurrence of leucophores on the caudal fin of the fish, *Oryzias latipes*, following administration of androgenic steroids. *Annot. Zool. Jap.*, 34, pp. 185–192, 1961.

73. Yamazaki, F. Sex control and manipulation in fish. *Aquaculture*, 33, pp. 329–354, 1983.

74. Jost, A. A new look at the mechanisms controlling sexual differentiation in mammals. *Johns Hopkins Med. J.*, 130, pp. 38–53, 1972.

75. C.D. Metcalfe and Y. Kiparissis. 1997. Biological responses of fish to flavonoid compounds. *Proc. of 18ᵗʰ Annual Meeting of Society of Environmental Toxicology and Chemistry*, Nov. 16–20, 1997, San Fransisco, CA, p. 280.

76. Denton, T.E., W.M. Howell, J.J. Allison, J. McCollum, and B. Marks. Masculinization of female mosquitofish by exposure to plant sterols and *Mycobacterium smegmatis. Bull. Environ. Contam. Toxicol.*, 35, pp. 627–632, 1985.

77. Howell, W.M. and T.E. Denton. Gonopodial morphogenesis in female mosquitofish, *Gambusia affinis affinis*, masculinized by exposure to degradation products from plant sterols. *Environ. Biol. Fish.*, 24, pp. 43–51, 1989.

78. Arcand-Hoy, L.D. and W.H. Benson. Fish reproduction: An ecologically relevant indicator of endocrine disruption. *Environ. Toxicol. Chem.*, 17, pp. 49–57, 1998.

79. Guillette, L.J., Jr., D.A. Crain, A.A. Rooney, and D.B. Pickford. Organization versus activation: The role of endocrine-disrupting contaminants (EDCs) during embryonic development in wildlife. *Environ. Health Persp.*, 103 (Suppl 7), pp. 157–164, 1995.

80. Gray, M.A., K.L. Teather, and C.D. Metcalfe. Reproductive success and behaviour of Japanese medaka (*oryzias latipes*) exposed to 4-*tert*-octylphenol. *Environ. Toxicol. Chem.* (Accepted).

81. MacLusky, N.J. and F. Naftolin. Sexual differentiation of the central nervous system. *Science*, 211, pp. 1294–1302, 1981.

Chapter 3

Assessment of the Impact of Environmental Mutagens and Genotoxins: An Overview

B.K. Burnison and Salem S. Rao

INTRODUCTION

The Canadian Environmental Protection Act (CEPA) is "an Act respecting the protection of the environment and human health." CEPA includes provisions dealing with toxic substances, nutrients, and the environmental effects operation. Setting priorities and accelerating the assessment of substances of greatest concern will be carried out in light of the latest scientific knowledge. Detection of environmental mutagens and genotoxins can be considered as the first step in the assessment of the impact of potential environmental carcinogenicity. Several studies have shown that effluent from bleached kraft mills (BKME) contain mutagenic substances.[1-5] These substances could be potentially harmful to fish and other aquatic biota exposed to mill effluents.[6,7] Some studies indicate that fish in the vicinity of pulp mill effluent discharges develop tumors[7,8] and other physiological and reproductive disorders.[9] No experimental evidence has directly linked these toxic effects to specific mutagenic or carcinogenic substances in BKME. However, Metcalfe et al.[10] observed that nonpolar compounds extracted from BKME had a weak promotional effect upon aflatoxin-initiated hepatocarcinogenesis in rainbow trout.

The Ames test, which measures mutagenic activity in histidine-deficient strains of *Salmonella typhimurium*,[11] has been used to determine the mutagenicity of complex environmental mixtures, including BKME. Previous studies have indicated that spent liquors from kraft mills cause mainly base-pair substitution mutations in the Ames test.[1,2,12-17] According to these studies, unconcentrated spent liquors from the first chlorination stages are mutagenic in *Salmonella* strains TA-1535 and TA-100, without metabolic activation.

Our earlier work[3] has shown that BKME contains direct-acting mutagens which give a positive response in the Ames test (fluctuation assay) with tester strain TA-100, without exogenous activation. The majority of mutagenic activity in BKME can be adsorbed onto XAD resins and eluted with polar solvents such as methanol or aqueous solutions of sodium hydroxide,[3,4] suggesting that the mutagens are polar compounds, and possibly weak organic acids. Recently, Burnison et al.[18] developed an extraction procedure using DEAE cellulose for concentrating active substances in pulp mill effluents. This procedure also appears to extract the majority of Ames mutagenic materials from mill effluent.[3]

Not all forms of DNA damage are detected in mutagenicity assays such as the Ames test,[19,20] so other in vivo and in vitro assays have been devised that test a spectrum of genotoxic effects. For example, Oda et al.,[19] developed the *Umu-C* genotoxicity assay using a *Salmonella typhimurium* recombinant strain TA-1535/ pSK1002, which detects genotoxic responses that increase expression of SOS repair activity associated with the *Umu-C* gene. This system has been used to screen for environmental mutagens and carcinogens.[21] Some Ames-positive substances are only mutagenic in prokaryotes, so other short-term assay systems with eukaryotes have been devised to detect DNA-damaging agents that may be carcinogenic to vertebrates.[22–24] Micronucleus assays detect chromosomal damage and breaks that lead to the formation of small fragments of chromatin outside of the main nucleus. Frequencies of micronuclei are increased in mammalian species and in fish by known clastogens.[25–27] DNA strand breaks can also be detected with various in vitro assays, such as the DNA precipitation assay[28] in which strand breaks generate soluble fragments of single strand DNA (ssDNA) that do not precipitate under the experimental conditions. The proportion of ssDNA in the supernatant in this procedure can be used as a screening assay for detection of direct-acting mutagens in environmental samples.

In this chapter we describe our study designed to determine whether Ames-positive mutagens isolated from BKME induce other types of genotoxic responses in in vitro and in vivo assays. The genotoxic responses to BKME extracts and semipurified fractions of the extracts were determined in the in vitro bacterial *Umu-C* assay, an in vivo hepatic micronucleus assay with rainbow trout, and in an in vitro assay for DNA strand breaks with mammalian cell lines.[29]

MATERIALS AND METHODS

Preparation of BKME Extracts

Bleached kraft mill effluent (BKME) was sampled prior to its discharge into the receiving river, collected in 200-L polyethylene barrels and transported to the laboratory by truck. The effluent source was a modernized kraft mill producing approximately 1100 tonnes of bleached pulp per day. The mill produces soft-wood and hardwood pulp on two separate pulping and bleaching lines. It employs extended oxygen delignification, high chlorine dioxide substitution, and biological treatment of effluent.[30,31] The effluent was processed according to the procedure shown in Figure 3.1.

The effluent (240 L) was pumped from the barrels with a Little Giant (March 3D) submersible pump into a continuous-flow centrifuge (Westfalia, 10,000 rpm) adjusted to a flow rate of 2 L×min^{-1}. The collected suspended solids (SS) were freeze-dried for later analyses. The supernatant was filtered through a 142 mm Gelman A/E (pore size, 1.0 μm) glass fiber filter by N_2 pressure (150 KPa). The filtrate was passed through a 0.2 μm microporous membrane (Pellicon® cassette GVLP, 5 ft^2 filtering area) housed in the Pellicon® cassette sanitary stainless cell (Millipore®). This tangential flow, ultrafiltration device allows the effluent to be pumped through the system by a Procon® rotary vane pump at a flow rate of 700 mL×min^{-1}. The >0.2 μm retentate (RET) was dialyzed with 4 L Milli-Q water before being removed from the tangential flow unit. The retentate was centri-fuged (5,500 g) and the supernatant frozen and freeze-dried. Samples of filtered (0.2 μm) mill effluent and effluent extracts were prepared for genotoxicity test-ing.

XAD adsorption: the macroreticular resins XAD-4 and XAD-8 (British Drug House) were thoroughly washed and allowed to settle in water to remove the "fines." The resins were placed separately into two-liter Soxhlet thimbles and extracted sequentially with methanol, acetonitrile, and dichloromethane (DCM). After the DCM had been drained, the resins were soaked in 1.0 N NaOH for 30 minutes and then washed with methanol. The cleaned resins were stored under methanol.

Approximately 500 g of cleaned resin was poured into glass chromatography columns (4.1 × 60 cm) which contained 200 mL of methanol. The XAD col-umns were washed with methanol to remove traces of DCM. The packed col-umns were subsequently washed with Milli-Q water, 1.0 N NaOH, and again with Milli-Q water until the pH was neutral.

The filtered effluent was pumped onto two XAD-4 columns, in series (initial and secondary), at 45 mL×min^{-1} with a peristaltic pump (Cole-Parmer Instru-ment Co.). The XAD-8 column was placed at the end of the series.

XAD elution: each column containing adsorbed BKME material was rinsed with Milli-Q water and then eluted with one-liter quantities of: (1) 0.05 N NaOH, rinsed with Milli-Q water; (2) methanol; and (3) DCM. The columns were drained and flushed with N_2 gas, to remove residual traces of polar solvents, then finally eluted with hexane. The DCM and hexane eluents were combined.

Fractions for bioassays: the NaOH fractions (NA4 or NA8) were neutralized to pH 7.2 with concentrated HCl and then frozen. The methanol extracts (ME4

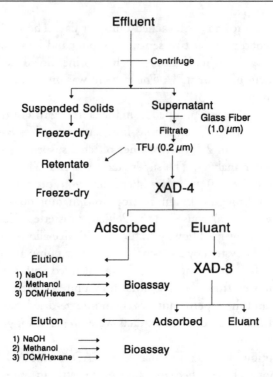

Figure 3.1. Experimental flow diagram for isolation of mutagenic fractions. From Rao, S.S., B.K. Burnison, D.A. Rokosh, and C.M. Taylor, *Chemosphere,* 28, 1859–1870, 1994. With permission.

or ME8) were concentrated by rotoevaporation under vacuum at 30°C and this methanol:water mixture was frozen. The DCM/hexane eluents were concentrated to about 10 mL and "washed" with hexane three times to remove traces of DCM. A precipitate formed at this stage for both XAD-4 and XAD-8 DCM/hexane eluents. This precipitate (HEP4 or HEP8) was removed by centrifugation (2,800 rpm for 10 min), then dissolved in dimethyl sulfoxide (DMSO). The centrifuged DCM/hexane fractions (HEX4 or HEX8) were recovered and 8 mL of DMSO was added to each. The remaining hexane was removed with a stream of N_2. The concentrated DCM/hexane eluent from XAD-4 also had a brown liquid (HEI4) that was immiscible in hexane. Traces of hexane were removed in a stream of N_2 gas from this HEI4 fraction. The freeze-dried fractions SS and RET were extracted with a mixture of acetone:hexane (1:1) to produce fractions SS-NP and RET-NP, respectively. A second aliquot of the SS and RET fraction was extracted with DMSO to produce fractions SS-P and RET-P, respectively. All ef-

Table 3.1. Fractions Collected for BKME Mutagens.[a]

Fraction	Sample	Volume (mL)	Concentration Factor
XAD-4			
NA4	NaOH eluent	2,000	120
ME4	Methanol eluent	250	960
HEX4	DCM/Hexane eluent	8.2	29,268
HEI4	Insoluble in hexane	2.5	96,000
HEP4	Particulate residue	8.0	30,000
XAD-8			
NA8	NaOH eluent	1,000	120
ME8	Methanol eluent	152	789
HEX8	DCM/Hexane eluent	2.0	60,000
HEP8	Particulate residue	4.1	29,268
Freeze-Dried			
SS-P	DMSO extract of SS	3.0	10,000
SS-NP	Hexane/acetone ext. of SS	2.0	10,000
RET-P	DMSO extract of RET	2.0	5,780
RET-NP	Hexane/acetone ext. of RET	2.0	10,000

[a] From Rao, S.S., B.K. Burnison, D.A. Rokosh, and C.M. Taylor, *Chemosphere,* 28, 1859–1870, 1994. With permission.

fluent fractions and the estimated concentration factors achieved in these fractions are listed in Table 3.1.

The resin adsorbents (XAD-4 and XAD-8) and the eluents (NaOH and MeOH) used to prepare various BKME extracts are summarized in Table 3.2.

The procedure[18] used to adsorb effluent mutagens on DEAE cellulose included centrifugation of pulp mill effluent at 10,000 g, followed by filtration of the supernatant through a 0.2 μm Pellicon cassette housed in a Millipore tangential flow apparatus. The filtrate was tested for genotoxic activity, and the remainder was adsorbed onto DEAE cellulose. Before use, the DEAE cellulose was rinsed with water and allowed to drain to dryness under vacuum. The mutagenic substances adsorbed to cellulose were extracted with methanol for one hour and then removed by glass fiber filtration. The cellulose was removed by filtration (glass fiber, 1.0 μm pore size) and the clean filtrates tested for mutagenicity/genotoxicity.

The majority of the BKME extracts that were tested in the genotoxicity assays were evaporated to dryness using rotary evaporator (30°C) to remove the elution solvent, and the residue was dissolved in Eagle Minimum Essential Medium (EMEM). However, not all samples were treated this way. The adsorbent,

Table 3.2. Protocols for Preparing BKME Extracts for Genotoxicity Tests.

Adsorbent	Eluent	Final Solvent	Genotoxicity Test
XAD-4	NaOH	0.05 mM NaOH	Ames[1]
XAD-4	NaOH	EMEM[2]	Ames, MN[3], Breaks[4]
XAD-4	Methanol	EMEM	Ames, Breaks
XAD-8	NaOH	EMEM	Ames, Breaks
XAD-8	Methanol	Water	Ames, Breaks
DEAE	Methanol	Methanol	Ames, Umu-C[5]

(1) Ames fluctuation assay, (2) Eagle's Minimal Essential Medium, (3) Fish hepatic micronucleus assay, (4) DNA strand breaks assay, (5) *Umu-C* assay. From Rao, S.S., B.A. Quinn, B.K. Burnison, M.A. Hayes and C.D. Metcalfe, *Chemosphere*, 31, pp. 3553–3566, 1995. With permission.

eluents and final solvents used to prepare all of the extracts are summarized in Table 3.2.

Bioassays

Seven different bioassays were used to test the toxicity, mutagenicity and genotoxicity of the effluent and effluent fractions: Microtox™ and *Daphnia magna* for acute toxicity; and the Ames test and SOS-Chromotest for genotoxicity.

Ames Salmonella Mutagenicity Assay

The Ames *Salmonella* mutagenicity test was conducted using the fluctuation assay protocol with the TA-100 tester strain and without exogenous activation (S-9). The fluctuation assay was conducted using a micro-well procedure and was scored by recording the number of wells containing revertant cells (positive wells).[32] Whole effluent samples and 1 μm filtrate fractions were tested for mutagenicity using the Ames fluctuation assay[32,33] at doses between 2.5 mL and 15 mL effluent. Effluent was filtered through 0.2 μm Teflon filters, to obtain bacteria-free samples, using a positive pressure with N_2 gas, to retain volatile compounds. Filtered BKME samples and extracts prepared from effluent (XAD-4, XAD-8, and DEAE) were tested for mutagenicity[32] at doses of 2.5 mL to 15 mL. All mutagenicity tests included a negative control and a positive control (sodium azide). With the fluctuation assay, a sample was considered mutagenic when one or more effluent doses (volumes) induced a significant ($p<0.05$) dose-related increase in the number of positive wells relative to the number of background mutations.[34] Ninety-six replicate wells were done per dose.

Effluent fractions were also assayed for mutagenicity using the Ames test, plate incorporation procedure[11] at doses of 25 to 200, which is equivalent to 3 mL to 1000 mL of original effluent. All mutagenicity tests included negative controls (solvent) and positive controls (sodium azide for strain TA100 and 2-nitrofluorene for strain TA98). The Ames plate incorporation procedure was scored by counting the number of colonies of revertant cells. A sample was considered mutagenic if it demonstrated a dose-related increase in the number of revertant colonies and with at least one dose of the sample inducing twice the number of revertants compared to the negative control (spontaneous revertants).

Umu-C *Bacterial Genotoxicity Assay*

For purposes of direct comparison with the bacterial genotoxicity assay *Umu-C* test, the DEAE concentrate was also tested for mutagenicity using the fluctuation assay protocol in serially diluted samples similar to those used in the *Umu-C* assay. Figure 3.2 summarizes the molecular basis of the *Umu-C* genotoxicity test. The *Umu-C* assay detects activation of the *Rec-A* operon (i.e., SOS response system) in a recombinant bacterial strain (*S. typhimurium* TA1535/pSK1002) through activation of the plasmid-associated *Umu-C* operon and downstream transcription of the *lacZ* gene, which directs the synthesis of β-galactosidase. The activity of β-galactosidase is detected with a colorimetric assay. The *Umu-C* genotoxicity test for detecting induction of the SOS response was carried out with the MeOH concentrate from DEAE cellulose, essentially as described previously by Oda et al.[19] The mutagen-induced DNA repair (SOS repair) has been proposed to be responsible for mutation caused by many DNA-damaging agents. In *E. coli,* the response of SOS repair involves several SOS genes including *Umu-C* and is regulated by *Lex*A and *rec*A. Based on the SOS repair, Quilardet et al.[24] were the first to describe a bacterial assay (SOS Chromotest) for detection of genotoxins by monitoring SOS responses. Using the same principle, of SOS response, and gene-fusion technique Oda et al.[19] constructed a tester strain *Salmonella typhimurium* containing plasmids which carry *Umu-C/Lac-Z* fusion gene. The sensitivity of *Umu-C* test has been shown to be similar to that of the Ames mutagenicity assay.[19]

Briefly, an overnight culture of the tester strain (TA-1535/pSK1002) in 10× tryptose-glucose-ampicillin (TGA) broth was exposed for 2 hours to sample material at different dilutions and tested for cytotoxicity. After 2 hours exposure, 30 mL of the organism/test solution mixture was transferred to 1× TGA broth and incubated for 2 hours at 37°C, and the cell density was read at 600 nm. After recording the growth response, the mixture (30 mL) was transferred to another multichannel plate containing lysis buffer to assess the β-galactosidase activity.

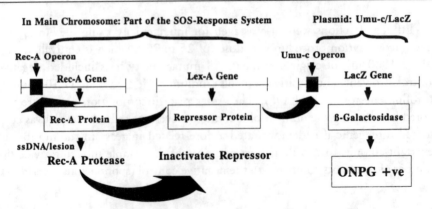

Figure 3.2. Molecular basis for the *Umu-C* genotoxicity assay using *Salmonella typhimurium* TA1535/pSK1002. From Rao, S.S., B.A. Quinn, B.K. Burnison, M.A. Hayes and C.D. Metcalfe, *Chemosphere*, 31, 3553–3566, 1995. With permission.

The level of β-galactosidase activity, which indicates activation of the *Umu-C* operon, is correlated with the amount of genotoxic material in the test medium.[35] A positive genotoxic response in the *Umu-C* assay occurs when there is an induction rate of at least 1.5 times the control. The growth factor indicated in Figure 3.2 indicates that the test substance was not toxic to the bacteria.

SOS-Chromotest

The SOS-Chromotest was performed on effluent samples and effluent sample fractions, both with and without exogenous activation (S-9). The SOS-Chromotest is a colorimetric assay using the bacterial strain *E. coli* K12-PQ37 which measures the induction of the SOS repair system and the coincident induction of the enzyme β-galactosidase.[36] A sample was considered genotoxic if it induced a dose-related increase in β-galactosidase activity.

Microtox Bioassay

The Microtox bioassay was performed using the procedure outlined by Dutka.[37] This bioassay measures toxicity by recording the dose-related decrease in bioluminescence of the marine bacterium *Photobacterium phosphoreum*. Each effluent and fraction sample was analyzed at dilutions of 45.4, 22.7, and 11.4% of the original sample. Results were expressed in terms of EC_{50}%.

Daphnia magna *Bioassay*

A 48-hour *Daphnia magna* test, using 10 organisms for each of the samples tested, was performed using the procedure outlined by Dutka.[37]

Fish Hepatic Micronucleus Assay

Rainbow trout, *Oncorhynchus mykiss* (25–30 g) were obtained from Rainbow Springs Hatchery in Thamesford, Ontario. After anasthesia with MS-222, five rainbow trout (for each treatment) were injected interperitoneally (i.p.) with either 2 mL of diethylnitrosamine (DEN) (0.05 mL/g) (positive control), 2.5 mL of water (negative control), or XAD-4/NaOH extract of effluent at doses of 1 mL (8.3 mL), 4 mL (33.3 mL), 7 mL (58.3 mL), 12 mL (100 mL), and 16 mL (133 mL) equivalents. The fish were then placed in 16–20 L aquaria containing 10 L of continuously-aerated, dechlorinated Burlington, Ontario tap water, and the water in each tank was replaced every 24 hours. After 48 hours, the fish were reinjected with the same volume of extract and placed back in the appropriate aquaria. After another 48 hours, the fish were injected a third time with the same volume of extract. At 48 hours following the final injection, the fish were injected with 2 mL of allyl formate (AF) to induce hepatocyte proliferation. AF induces liver necrosis and subsequent regenerative proliferation of hepatocytes in the liver.[27] During the next 12 hours, mortalities occurred among the AF-injected fish; possibly because of the trauma of the experimental procedure. Ninety-six hours after the AF injection, the remaining fish were sacrificed with a blow to the head and severing of the spinal column.

Hepatocyte suspensions were prepared from trout livers as described by Williams and Metcalfe.[27] Briefly, the livers were macerated and rinsed with a 1% citrate solution, and then placed in 8 mL of 0.1% collagenase and vortexed approximately 8 times in a 30 minute period. The hepatocyte suspension produced by this treatment was centrifuged for 3–4 minute, and the hepatocyte pellet was resuspended in 2–3 drops of citrate solution. The suspension was applied to microscope slide with a Pasteur pipet[27] and air dried. The slides were placed in Carnoy's fixative for 1 hour, and then allowed to dry. Slides were stained with Schiff's reagent and counter-stained with 2% light green in ethanol. Slides were examined by a light microscope with an oil immersion lens (1,000×), and micronuclei were identified as Schiff-positive microbodies in the cytoplasm. All slides were coded and scored blind and 1,000 hepatocytes were scored per fish. Differences between the mean incidence of micronuclei (per 1,000 hepatocytes) in each treatment were tested statistically using nonparametric methods (Kruskal-Wallis test).

Mammalian Cell DNA
Precipitation Assay

Chinese hamster lung V79-4 cells, mouse normal liver T1B-73 cells, and mouse transformed liver T1B-75 cells (all from ATCC) were grown in 75 cm² culture flasks in Eagles minimum essential medium (EMEM), pH 7.0, supplemented with 10% (v/v) fetal calf serum (GIBCO), 472 units of penicillin, and 94 mg of streptomycin/mL. The cells were kept in humidified CO_2/air (1:19) at 37°C and maintained in exponential growth as a monolayer by subcultivating biweekly. Cells were harvested with 0.1% trypsin and resuspended at a density of 2×10^5 cells/mL of supplemented EMEM, containing 1 mCi/mL ³H-thymidine (84 Ci/mM). To each well of a 12-well culture plate (3.8 cm²/22 mm dia. well), 1 mL of cell suspension was added and the cells incubated for 20 hours, followed by 2-hour incubation in nonradioactive medium. Test fractions were added to the wells and the cells incubated for an additional 1 hour. The culture medium was removed and the plates placed on ice. Indirect-acting genotoxic activity in these fractions were examined by the addition of washed rat liver S-9 microsomes (1 mg/mL microsomal protein). An attempt was also made to show glutathione S-transferase-dependant protection against the mutagenic fractions by the addition of 1 mg/mL of dialyzed liver cytosol from rat, rainbow trout, white sucker (*Catostomus commersoni*), and human liver in the presence of 5 mM reduced glutathione. Dialyzed cytosol without glutathione served as a control to demonstrate GST-specific activity.

A modified version of the DNA precipitation assay described by Olive[28] was used to detect ssDNA breaks. To each culture well, 0.5 mL of lysis buffer (10 mM Tris, 10 mM EDTA, 0.05 M NaOH, 2% SDS, pH 12.4), was added. The cells were lyzed for 1 minute, then carefully transferred to a centrifuge tube to which 0.5 mL KCl was added. The tubes were placed in a water bath at 65°C for 10 minutes, then cooled on ice for 10 minutes. Low molecular weight soluble ssDNA was released from the bulk DNA-protein complex during centrifugation at 3,500 rpm for 10 minutes at 4°C. The supernatant was decanted into a scintillation vial and neutralized with the addition of 1 mL of 0.5 N HCl. The pellet was solubilized in 2 mL of distilled water at 65°C for 10 minutes, and decanted into a scintillation vial. To each vial, 10 mL of scintillation cocktail (Cytoszint, ICN Biochemicals) was added and samples counted in a scintillation counter. Results are presented as percentage precipitation of DNA in comparison with vehicle-treated control cultures (i.e., no extract added). The percentage is calculated for each sample and normalized to the values of precipitated DNA in untreated controls, taken as 100%. The exposure experiments were performed in triplicates for each dose.

RESULTS AND DISCUSSIONS

Genotoxicity of Effluent and Effluent Fractions

As shown in Table 3.3, the mutagenic activity was detected in the mill effluent. Using the Ames fluctuation assay, a procedure which allowed the incorporation of sample volumes as large as 15 mL into the assay medium. This BKME
effluent at volumes of 15 mL contained sufficient levels of mutagenic material to
be detected by the fluctuation assay. Samples taken six months later at the same
site confirmed the presence of Ames mutagenic substances (minimum volume
needed was 10 mL). Mutagenic substances present in the chlorination stage effluents are primarily direct acting mutagens (not requiring activation) and detected
with strain TA-100.[2,38] In our studies, mutagenicity of BKME samples was assessed using strain TA-100 and the mutagenic components consisted of direct
acting mutagens, since exogenous activation (S-9) was not performed. Because of
the possibility of bacterial contamination, effluent samples were filtered through
0.2 μm membrane filters. This procedure, however, did not remove mutagens, as
subsequent testing of effluent fractions indicated the majority of activity was associated with the water-soluble filtrate, since both suspended solids centrifuged
from the sample (SS-P and SS-NP) and materials retained after ultrafiltration of
the sample (RET-P and RET-NP) were not mutagenic.

Testing of subsequent fractions of the aqueous phase of BKME demonstrated
the mutagenic material was retained on XAD-4 and XAD-8 resins. Mutagenic
compounds retained on XAD-4 and XAD-8 resins were efficiently eluted with
NaOH or methanol as indicated by mutagenicity data for the various fractions
(Table 3.3). Concentration of mutagenic compounds in these fractions, were
sufficiently high that the Ames plate incorporation assay could measure mutagenic
activity. The relative low numbers of the revertant colonies (compared to background values) obtained in the 2.5 and 5.0 mL effluent equivalent values may be
due to the normal variations encountered.

As the levels of mutagenic components increase, a dose-related type of response is seen (Table 3.3). However, the DCM/hexane eluent of materials retained on the XAD-4 was only weakly mutagenic, and DCM/hexane eluent of
XAD-8 resin was nonmutagenic (Table 3.3). Since NaOH and methanol elute
polar materials from XAD resins, these data suggest the mutagenicity was associated with the polar organic materials in this BKME sample.

The results of the Ames test were used to estimate the relative potencies of
various BKME fractions. Mutagenic potency was defined here as the minimum
volume of a fraction, equivalent to the volume of final effluent, in which an
increase in mutagenic response was first detected. Based on this assessment, BKME
material eluted by NaOH from both XAD-4 and XAD-8 had the highest mu-

Table 3.3. Mutagenicity of Effluent and Effluent Fractions.[a,b,c]

Effluent Dose (mL)	Effluent Nonfiltered	<0.2 μm Filtrate	Dose of Fraction (μL)	NA4	ME4	HEX4	HEP4	NA8	ME8
	Fluctuation Assay (Number Positive Wells)			Plate Incorporation (Number Revertants Per Plate)					
				XAD-4 Resin				XAD-8 Resin	
0	13	13	0	186	186	143	143	186	186
2.5	14	13	25	135	103	140	145	134	97
5.0	14	16	0	138	132	178[e]	176[e]	140	124
10.0	16	20	75	320[e]	236[e]			401[e]	146
15.0	26[d]	39[d]	100	359	380	180	194	446	240[e]
			200	359	410			499	510

[a] From Rao, S.S., B.K. Burnison, D.A. Rokosh, and C.M. Taylor, *Chemosphere*, 28, pp. 1859–1870, 1994. With permission.

[b] Mutagenicity was not detected in effluent and fractions with strain TA 98.

[c] Remaining fractions tested (see Table 3.1) were nonmutagenic with strain TA 100.

[d] Number of positive wells significantly (p<0.05) greater than control (0 ml dose).

[e] Number of revertants considered greater than spontaneous background mutation (0 μL dose).

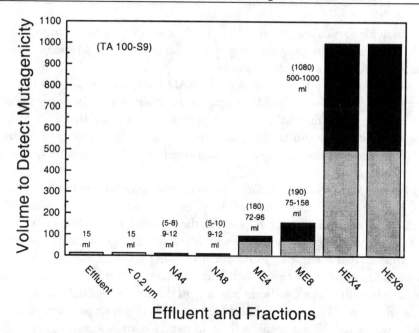

Figure 3.3. Mutagenic potencies of effluent fractions. Bars (upper and lower range) represent the *minimum* volume of effluent needed to detect mutagenicity. From Rao, S.S., B.K. Burnison, D.A. Rokosh, and C.M. Taylor, *Chemosphere*, 28, 1859–1870, 1994. With permission.

tagenic potencies (Figure 3.3). Mutagenicity in NaOH eluents of both XAD resins was detected at volumes equivalent to between 9 and 12 mL of the whole effluent. Potencies of these NaOH eluents were similar to those of the whole effluent, where mutagenicity was detected at a volume of 15 mL. Assuming that the fluctuation assay and the plate incorporation assay are comparable, then the mutagenicity of the XAD-4 and XAD-8 concentrates represent approximately 75% of the total mutagenic activity detected in the pulp mill effluent prior to its discharge into the receiving water. Methanol eluents of both XAD-4 and XAD-8 resin were also mutagenic but they were less potent than NaOH eluents. Methanol eluents from the XAD-4 resin were slightly more mutagenic than that of XAD-8 (Figure 3.3). The activity of the XAD-4 methanol eluent was detected at an equivalent volume of 72 to 96 mL compared to 75 to 158 mL for the XAD-8 methanol fraction.

Nonpolar BKME material, adsorbed to the resins and eluted with DCM/ hexane, was either weakly mutagenic or nonmutagenic. A slight increase in revertants was measured in XAD-4 DCM/hexane eluents only, where the mutagenic potency was equivalent to 500 to 1000 mL of whole effluent (Figure 3.3). Hexane-insoluble particulate material eluted by these nonpolar solvents from XAD-4 (HEP4) contained mutagenic activity at a similar potency. DCM/hexane eluents from XAD-8 resins were nonmutagenic.

Results of mutagenicity testing of the XAD fractions indicate that the mutagenicity measured in this BKME sample was associated with polar (water soluble) and possibly acidic materials. Strongest potencies were associated with NaOH eluents, which likely contain organic acids. Moreover, the majority of mutagenicity measured in the whole effluent was largely associated with these NaOH eluted materials.

Figure 3.4 shows the distribution of Ames mutagenic activity in filtered (<0.2 μm) effluent (Figure 3.4B) and the DEAE extract from effluent (Figure 3.4C). Both sets of data show a dose-response in the fluctuation assay, which indicates that the DEAE extraction method effectively extracts mutagenic compounds from the effluent. It is interesting to note that the dose at which the mutagenic response is significant in the assay with filtered effluent (2.5 mL) approximates the dose at which there is a significant response in the assay with DEAE extract (3.1 mL equivalents). When the filtered effluent material that was not adsorbed by the DEAE was tested for mutagenic activity in the fluctuation assay (Figure 3.4D), there was some residual mutagenic activity, but only at a dose of 15 mL.

The Ames mutagenicity and *Umu-C* genotoxicity were measured in the MeOH extract of the DEAE adsorbed materials. The assay results (Figure 3.4A) indicate that this extract also induced a dose-dependent SOS repair response in the *Umu-C* assay, but this genotoxic response reached a level of significance at a dose of 0.28 mL equivalents, which indicates that this assay is more sensitive than the fluctuation assay by an order of magnitude. However, the responses of the bioassay organisms to the extract in the two protocols were comparable. Based on the present study, it seems that these two assays are complementary and practical *in vitro* tests for screening complex environmental mixtures for genotoxic potential.

Incidence of Hepatic Micronuclei in Trout

There was an elevated incidence of micronucleated hepatocytes relative to controls in rainbow trout injected with XAD-4/NaOH extract of mill effluent at a dose of 4 mL equivalents (Figure 3.5). However, the trout in the positive control (DEN) treatment did not have statistically significant increase in hepatic micronuclei (MN). There is no apparent explanation for the lack of response to DEN since this indirect-acting genotoxic compound has been shown previously to induce micronuclei in rainbow trout.[27]

Effluent extract at a dose of 1 mL equivalent indicated twice as great a genotoxic effects as positive controls. A mean MN incidence of 13 micronuclei per 1,000 hepatocytes was observed at the 4 mL equivalent dose, but higher doses (7, 12,

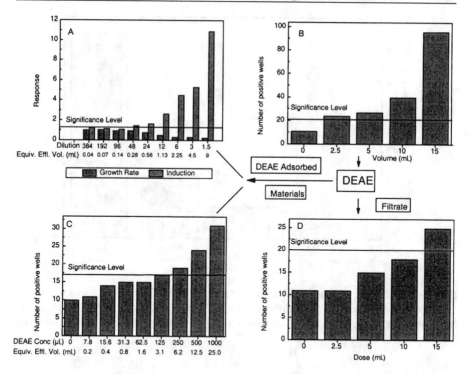

Figure 3.4. Genotoxic responses in Ames fluctuation and *Umu-C* assays induced by pulp mill effluent and their fractions. (A) *Umu-C* genotoxicity assay with DEAE methanol extract; (B) Ames fluctuation assay with < 0.2 μm filtered effluent; (C) Ames fluctuation assay with DEAE methanol extract; and (D) Ames fluctuation assay with DEAE filtrate. From Rao, S.S., B.A. Quinn, B.K. Burnison, M.A. Hayes, and C.D. Metcalfe, *Chemosphere*, 31, 3553–3566, 1995. With permission.

and 16 mL equivalents) were toxic to the experimental fish. It is possible that the MN observed in the 4 mL equivalent treatment were caused through either inhibition of chromosomal segregation in dividing cells through the activity of chemical "spindle poisons" in the extract, or alternatively, were caused by chemically-induced apoptosis of hepatocytes.[39] These latter mechanisms do not involve genotoxic responses. However, this positive response in the hepatic MN assay with trout is consistent with previous observations of elevated numbers of micronuclei in the peripheral erythrocytes of fish exposed to pulp mill effluent.[26]

DNA Damaging Activity in Mammalian Cells

Figure 3.6 shows the DNA damaging effects of the XAD-4/MeOH extract, in the exponentially growing Chinese hamster lung V79-4 cells, mouse liver T1B-

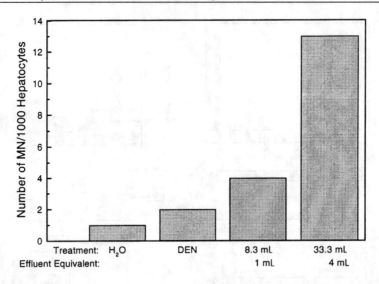

Figure 3.5. Effect of pulp mill mutagens (XAD-4 NaOH) on the induction of hepatocyte micronuclei in rainbow trout. From Rao, S.S., B.A. Quinn, B.K. Burnison, M.A. Hayes, and C.D. Metcalfe, *Chemosphere*, 31, 3553–3566, 1995. With permission.

73 cells, and transformed liver T1B-75 cells exposed to XAD-4/MeOH extract for 1 hour. A dose-related release of soluble DNA fragments (putatively due to single or double strand DNA breaks) was detected for all the mammalian cells used in these studies.

In addition, a dose response was detected for V79-4 cells exposed to XAD-4/ NaOH extract (not shown), which were the most sensitive cells used. Addition of liver cytosol (± reduced GSH) from various species did not modify the activity of XAD-4/MeOH extract in V79 cells (Table 3.4). Also, rat microsomes did not substantially alter the response to XAD-4/MeOH extract in all 3 cell lines (Table 3.5). The XAD-8/NaOH extract at doses up to 1000 mL had no activity in the DNA precipitation assay with V79, T1B-73, and T1B-75 cells, in the presence or absence of rat microsomes. The fraction extracted by XAD-8 and eluted with NaOH, and extracted by XAD-8 and eluted with MeOH did not induce DNA damage in assays with V79 cells.

While mutagenic potency of XAD-4/NaOH extract was detected at doses equivalent to approximately 9–12 mL of effluent,[3] significant responses to the extract were observed in this assay for DNA strand breaks at doses of approximately 5–10 mL equivalents. The above data indicate that the genotoxic potential of weakly acidic compounds extracted using the XAD-4 resin, and illustrate the usefulness of these in vitro methods for assessing the environmental significance of pulp mill mutagens.

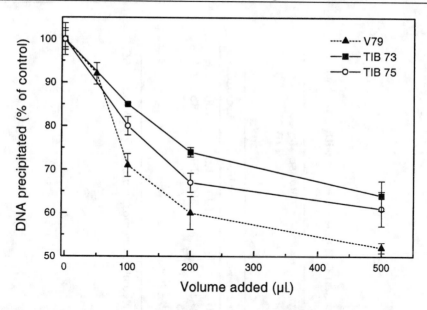

Figure 3.6. XAD-4 methanol extract induction of DNA strand breaks in various cell lines; one microliter extract equals one milliliter BKME. From Rao, S.S., B.A. Quinn, B.K. Burnison, M.A. Hayes, and C.D. Metcalfe, *Chemosphere*, 31, 3553–3566, 1995. With permission.

Table 3.4. Comparison of Effects of Added Liver Cytosol +/– Reduced Glutathione on DNA Strand Breaks in V79 Cells Exposed to 100 mL of XAD-4 MeOH Extract.[a]

Preparation	Cytosol ± GSH	% DNA Breaks[b]
V-79 cells+	+	79 ± 1.0
XAD-4-MeOH extract	–	79 ± 2.3
V-79 cells+		
XAD-4-MeOH+		
liver cytosol from:		
a) Rat	+	76 ± 1.0
	–	77 ± 1.4
b) Sucker	+	81 ± 1.3
	–	80 ± 1.1
c) Trout	+	79 ± 3.1
	–	79 ± 1.8
d) Human	+	80 ± 0.6
	–	80 ± 1.3

[a] From Rao, S.S., B.A. Quinn, B.K. Burnison, M.A. Hayes, and C.D. Metcalfe, *Chemosphere*, 31, pp. 3553–3566, 1995. With permission.
[b] The percentage is calculated for each sample and normalized to the value of precipitated DNA in untreated controls, taken as 100%.

Table 3.5. Comparison of DNA Precipitation from Three Mammalian Cell Lines Treated with XAD-4 MeOH and the Effect of Added Rat Hepatic Microsomes.[a]

XAD-4 MeOH Volume Added[b] (μL)	Chinese Hamster Fibroblasts V79 Modifier		Mouse Hepatocytes T1B-73 Modifier		Mouse Hepatocytes T1B-75 Modifier	
	Control	Microsome	Control	Microsome	Control	Microsome
0	100[a] ± 3.7	100 ± 1.6	100 ± 2.5	100 ± 1.0	100 ± 1.9	100 ± 2.6
50	92 ± 2.5	96 ± 1.4	85 ± 0.3	89 ± 1.7	80 ± 2.1	83 ± 1.4
100	71 ± 0.5	76 ± 1.2	74 ± 1.1	83 ± 2.0	67 ± 2.2	73 ± 2.6
200	59 ± 1.1	69 ± 1.6	64 ± 3.4	61 ± 4.0		
500	52 ± 1.2					

[a] From Rao, S.S., B.A. Quinn, B.K. Burnison, M.A. Hayes and C.D. Metcalfe, *Chemosphere*, 31, pp. 3553–3566, 1995. With permission.

[b] Effluent Equivalency 1 mL = 1000 mL.

In conclusion, although some of these bioassays are relatively more sensitive than the others, this study has demonstrated that application of the concept of a battery of mutagenicity and genotoxicity tests would enhance the overall detection and assessment of the impact of environmental mutagens and genotoxins.

SOS-Chromotest

Results of SOS-Chromotest bioassays of the original BKME sample indicated that there was mutagenic activity, requiring exogenous activation (S-9) in the original unfiltered effluent sample. However, this activity could not be confirmed in particulate materials (SS and RET) recovered from this effluent. Moreover, it was expected that the SOS-Chromotest would not detect mutagenicity in the whole effluent without the exogenous activation with S9, since 15 mL of effluent was required to detect activity in the fluctuation assay.

Results of the SOS-Chromotest on NaOH and methanol eluents of XAD adsorbed materials confirmed the results of the Ames test on these fractions (Table 3.6). Genotoxicity was detected in NaOH and methanol eluents from both XAD-4 and XAD-8 resins. However, it was not possible to calculate the potencies in these fractions from the SOS-Chromotest data. The SOS-Chromotest, with S-9 activation, detected genotoxicity in the DCM/hexane fractions (HEX-4, HEX-8, and HEX-8 particulate). These results were not confirmed with the Ames test, since S-9 activation was not employed with this test. However, these SOS-Chromotest data with DCM/hexane fractions indicate that nonpolar, promutagenic compounds may also be present in this BKME sample.

Detection of mutagenicity in BKME effluents confirms the findings of previous studies[2,16] in which mutagenicity was also detected using strain TA 100 without exogenous activation. However, although effluents gave a positive response in in vitro assays, such as the Ames test, genotoxicity could not be demonstrated in mammalian in vivo assays.[2] Our current studies also demonstrated activity with an in vitro assay.

Our data confirms those of Metcalfe et al.,[4,10] who demonstrated retention of polar mutagenic compounds from kraft mill effluents on XAD-7 resin. However, our study demonstrated that mutagenic compounds could also be retained on XAD-4 and eluted with polar solvents, in contrast to the use of a nonpolar elution solvent (diethyl ether) used by the previous authors. The compounds in the effluent which account for the mutagenic activity in the Ames test are unknown. However, the mutagenic materials were polar in nature and may be organic acids (resin acids), since they were eluted from the XAD resins with sodium hydroxide.

Holmbom et al.[1] described a compound, 3-chloro-4-(dichloromethyl)-5-hydroxy-2(5H)-furanone ("MX"), which could be the most mutagenic compound in pulp mill effluents. MX was shown to be neutralized by sulfur compounds such as glutathione and sulfite, which probably accounts for the inability to detect mutagenicity of BKME effluents in in vivo mammalian assays.[38] Horth et al.[40] showed with chlorinated tyrosine solutions and with drinking water that MX was not retained by XAD-2 when the sample was at pH 6.2. Our samples had a higher pH of 7.3 and it is doubtful that any MX, if present, would be concentrated on our XAD-4 and XAD-8 resins. In addition other factors, such as the instability of MX at high pH[38,41] and in DMSO,[40] would negate its presence in our fractions. Nonpolar mutagenic substances, such as chloroacetones and chloropropenes, have been isolated from BKME.[2] These compounds, if present in our effluent samples, would be recovered in the HEX4 fraction. However, the mutagenic potency of HEX4 was very weak, or not detectable. Metcalfe et al.[4] found that nonpolar compounds from BKME enhanced the carcinogenic response to aflatoxin B_1, but were not carcinogenic themselves.

Acute Toxicity Bioassays

Two acute toxicity assays were employed in this study, the Microtox and *Daphnia magna* bioassay.[37] Summaries of the results of these tests are shown in Table 3.6. Whole effluent samples were nontoxic with the Microtox bioassay, but unfortunately were not tested with the *Daphnia magna* bioassay. Microtox bioassay data, although limited, indicate that this effluent sample likely did not have an acutely toxic impact on the receiving water.

Concentrated effluent fractions were tested for toxicity to determine the characteristics of toxic substances in this effluent, even though this toxicity was considered to be less than those having an environmental impact. With the Microtox bioassays, both the XAD-4 and XAD-8 adsorbed materials, eluted with NaOH or methanol, were toxic. Expressed in terms of EC_{50} values, the XAD-methanol extracts (EC_{50} 19.76% for ME4, EC_{50} 33.85% for ME8) were more toxic than NaOH fractions (EC_{50} 73.13% for NA4, EC_{50} 56.94% for NA8). However, materials extracted from effluent suspended particulates (SS-P, SS-NP) or retained by ultrafiltration of the effluent (RET-P and RET-NP) were considerably more toxic than the XAD adsorbed materials.

With *Daphnia magna*, different toxic responses were measured. Toxicity was only detected in undiluted NA4 fraction (80% survival) and with the ME8 fraction (70% survival). However, the toxicity of XAD fractions, as measured by the two toxicity bioassays, did not correlate. Still, XAD-adsorbed effluent fractions shown to be toxic were those which were also the most mutagenic. Materials responsible for increased toxicity may also be responsible for mutagenic/genotoxic responses. More detailed chemical analyses would be necessary to resolve this supposition. Moreover, since the original effluent sample was nontoxic in these

Table 3.6. Toxicity and Mutagenicity of Effluent and Its Fractions.[a]

		Toxicity		Genotoxicity			
				SOS Chromotest[b]		Ames Test (TA100)	
Extract	Conc. Factor	Microtox[b] EC$_{50}$%	D. magna Undiluted	Response	Range (%)	Response	Equivalent Volume (mL)
Whole effluent		Negative	NP[d]	S9[f]	3.13	NP	NA[e]
Effluent:							
0.2 μm filtered		NP	NP	NP	NA	Positive	15
After centrifugation		Negative	NP	NP	NA	NP	NA
After GF (>0.2 μm)		Negative	NP	Negative	NA	Positive	15
XAD-4 NaOH	12x	73.13	EC$_{20}$	Positive	6.25–100	Positive	9–12
XAD-4 MeOH	960x	19.76	Negative	Positive	6.25–100		72–96
HEX-4	1000x	Negative[c]	NP	S9	3.13 Positive	Slightly	500–1000
HEI-4 (insoluble)	1000x	Negative[c]	NP	Negative	NA	Negative	ND[g]
HEP-4 (particulate)	1000x	62.37[c]	NP	Negative	NA	Slightly Positive	500–1000
XAD-4 H$_2$O rinse		Negative	NP	Negative	NA	Negative	ND
XAD-8 NaOH	1200x	56.94	Negative	Positive	6.25–100	Positive	9–12
XAD-8 MeOH	789x	33.85	EC$_{50}$	Positive	12.25–100	Positive	79–158
HEX-8	1000x	Negative[c]	NP	S9	0.1	Negative	ND
HEP-8 (particulate)	1000x	Negative[c]	NP	S9	50	Negative	ND
XAD-8 H$_2$O rinse		Negative	NP	Negative	NA	Negative	ND
Freeze-dried sample		NP	NP	NP	NA	Negative	ND

Table 3.6. Toxicity and Mutagenicity of Effluent and Its Fractions[a] (continued).

| Extract | Conc. Factor | Toxicity | | Genotoxicity | | | |
| | | Microtox[b] EC$_{50}$% | D. magna Undiluted | SOS Chromotest[b] | | Ames Test (TA100) | |
				Response	Range (%)	Response	Equivalent Volume (mL)
SS-P		2.72	NP	Negative	NA	Negative	ND
SS-NP		0.05	NP	Negative	NA	Negative	ND
RET-P		7.65	NP	Negative	NA	Negative	ND
RET-NP		14.16	NP	Negative	NA	Negative	ND

[a] From Rao, S.S., B.K. Burnison, D.A. Rokosh, and C.M. Taylor, *Chemosphere*, 28, pp. 1859–1870, 1994. With permission.

[b] Samples tested at 1% sample concentration.

[c] Samples tested at 0.9% due to toxicity of blank control.

[d] NP = test not performed.

[e] NA = nonapplicable, as test was not performed.

[f] S9 = Only positive with S9 activation.

[g] ND = not detected.

bioassays, the significance of the results of the concentrated in effluent fraction produced in this study need to be reassessed.

ACKNOWLEDGMENTS

Many people provided valuable advice and assistance for this chapter. The authors would like particularly to express their sincere thanks to Drs. P.V. Hodson and J. Carey for their support and encouragement throughout this investigation. The *Umu-C* assay was performed at the Technical University of Berlin during a visit by SSR to Germany under the Canada-Germany Bilateral Agreement Project.

REFERENCES

1. Holmbom, B.R., R.H. Voss, R.D. Mortimer, and A. Wong. Isolation and identification of an Ames-mutagenic compound present in kraft chlorination effluents. *Tappi,* 64, pp. 172–174, 1981.
2. Douglas, G.R., E.R. Nestman, A.B. McKauge, O.P. Kamra, E.G.-H. Lee, J.A. Ellenton, R. Bell, D.J. Kowbel, V. Liu, and J. Pooley. Mutagenicity of Pulp and Paper Effluent: A Comprehensive Study of Complex Mixtures, in *Application of Short-Term Bioassays in the Analysis of Complex Environmental Mixtures III,* Waters, M., S. Sandhu, L. Claxton, J. Lewtas, S. Nestow, and N. Chernoff, Eds., Plenum, New York, 1983, pp. 431–460.
3. Rao, S.S., B.K. Burnison, D.A. Rokosh, and C.M. Taylor. Mutagenicity and toxicity assessment of pulp mill effluent. *Chemosphere,* 28, pp. 1859–1870, 1994.
4. Metcalfe, C.D., M.E. Nanni, and N.M. Scully. Carcinogenicity Testing of Bleached Kraft Mill Effluent Using *In Vivo* and *In Vitro* Assays, in *Environmental Research: Proceedings of the Technology Transfer Conference, Toronto, ON, 1991,* 1991, p. 267.
5. Munkittrick, K.R., C. Portt, C.J. Van Der Kraak, I. Smith, and D.A. Rokosh. Impact of bleached kraft mill effluent on liver MFO activity, serum steroid levels and population characteristics of Lake Superior white sucker population. *Can. J. Fish. Aquat. Sci.,* 48, pp. 1371–1380, 1991.
6. Hodson, P.V., M. McWhitter, K. Ralph, B. Grey, D. Thivierge, J.H. Carey, G. Van Der Kraak, D.M. Whittle, and M.C. Levesque. Effects of bleached kraft mill effluent on fish in the St. Maurice river, Quebec. *Environ. Toxicol. Chem.* 11, pp. 1635–1651, 1992.
7. Smith, I. and D.A. Rokosh. An Epidemiological Study of a White Sucker (*Catastomus commersoni*) Population Inhabiting the Kaministiquia River, Thunder Bay, Ontario, in *Proceedings 10th Annual Meeting, Society of Environmental Toxicology and Chemistry, Transboundary Pollution, Toronto, ON,* 1989, pp. 262–265.

8. Kinae, N., M. Yamashita, I. Tomita, I. Kimura, H. Ishida, H. Kumai, and G. Nakamura. A possible correlation between environmental chemicals and pigment cell neoplasia in fish. *Sci. Total Environ.*, 94, pp. 143–153, 1990.

9. Munkittrick, K.R., G.J. Van Der Kraak, M.E. McMaster, and C.B. Portt. Response of hepatic MFO activity and plasma sex steroids to secondary treatment of bleached kraft mill effluent and mill shutdown. *Environ. Toxicol. Chem.*, 11, pp. 1427–1439, 1992.

10. Metcalfe, C.D., M.E. Nanni, and N.M. Scully. Carcinogenicity and mutagenicity testing of extracts from bleached kraft mill effluent. *Chemosphere*, 30, pp. 1085–1095, 1995.

11. Maron, D. and B.N. Ames. Revised methods for the Salmonella mutagenicity test. *Mutation Res.*, 113, pp. 173–215, 1983.

12. Ander, P., K.E. Eriksson, M.C. Kolar, K. Kringstad, U. Rannug, and C. Ramel. Studies on the mutagenic properties of bleaching effluent, *Svensk Papperstidn.*, 80, pp. 454–459, 1977.

13. Bjorseth, A., G.E. Carlberg, and M. Moller. Determination of halogenated organic compounds and mutagenicity testing of spent bleach liquors. *Sci. Total Environ.*, 11, pp. 191–211, 1979.

14. Eriksson, K.E., M.C. Kolar, and K. Kringstad. Studies on the mutagenic properties of bleaching effluent, Part 2, *Svensk Papperstidn.*, 82, pp. 95–104, 1979.

15. Rannug, U., D. Jenssen, C. Ramel, K.E. Eriksson, and K. Kringstad. Mutagenic effects of effluent from chlorine bleaching of pulp. *J. Toxicol. Environ. Health*, 7, pp. 33–47, 1981.

16. Holmbom, B., R.H. Voss, R.D. Mortimer, and A. Wong. Fractionation, isolation, and characterization of Ames mutagenic compounds in kraft chlorination effluent. *Environ. Sci. Technol.*, 18, pp. 333–337, 1984.

17. Monarca, S., J.K. Hongslo, A. Kringstad, and G.E. Carlberg. Mutagenicity and organic halogen determination in body fluids and tissues of rats treated with drinking water and pulp mill bleachery effluent concentrates. *Chemosphere*, 13, pp. 1271–1281, 1984.

18. Burnison, B.K., P.V. Hodson, D. Nuttley, and S.M. Efler. Isolation and characterization of an MFO inducing fraction from BKME. *Environ. Chem. Toxicol.*, 15, pp. 1524–1531, 1996.

19. Oda, Y., S. Nakamura, I. Oki, T. Kato, and H. Shinagawa. Evaluation of the new system (*Umu-test*) for the detection of environmental mutagens and carcinogens. *Mutation Res.*, 147, pp. 219–229, 1985.

20. Setlow, R.B., F.F. Ahmed, and E. Grist. *Xeroderma pigmentosum*, pp. Damage to DNA is Involved in Carcinogenesis, in *Origins of Human Cancer*, Hiatt, H.H., J.D. Watson, and J.A. Winsten, Eds., Cold Spring Harbor Laboratory, Cold Spring Harbor, New York, 1977, pp. 889–902.

21. Kato, T. and Y. Shinoura. Isolation and characterization of mutants of *E. coli* deficient in induction of mutations by ultraviolet light. *Mol. Gen. Genet.*, 156, pp. 121–131, 1977.

22. Kada, T., K. Tutikawa, and Y. Sadie. In-vitro and host mediated "rec-assay" procedures for screening chemical mutagens; and phloxone, a mutagenic red dye detected. *Mutation Res.*, 16, pp. 165–174, 1972.

23. Ames, B.N., J. McCann, and E. Yamasaki. Methods for detecting carcinogens and mutagens with Salmonella/mammalian microsome mutagenicity test. *Mutation Res.*, 31 pp. 347–364, 1975.

24. Quilardet, P., C. de Bellecombe, and M. Hofnung. The SOS chromotest, a colorimetric bacterial assay for genotoxins: validation study with 83 compounds. *Mutation Res.*, 147, pp. 79–95, 1985.

25. Cliet, I., E. Fournier, C. Melcion, and A. Cordier. *In vivo* micronucleus test using mouse hepatocytes. *Mutation Res.*, 216, pp. 321–326, 1989.

26. Das, R.K. and Nanda, N.K. Induction of micronuclei in peripheral erythrocyte of fish *Heteropneustes fossilis* by mitomycin C and paper mill effluent. *Mutation Res.*, 175, pp. 67–71, 1986.

27. Williams, R.C. and C.D. Metcalfe. Development of an in vivo hepatic micronucleus assay with rainbow trout. *Aquat. Toxicol.*, 23, pp. 193–202, 1992.

28. Olive, P.L. DNA Precipitation Assay: A rapid and simple method for detecting DNA damage in mammalian cells. *Environ. Mol. Mutagenesis*, 11, pp. 487–495, 1988.

29. Rao, S.S., B.K. Burnison, S. Efler, D.A. Rokosh, E. Wittekindt, and P.D. Hansen. Assessment of the genotoxic potential of pulp mill effluent and an effluent fraction using the Ames mutagenicity and the *Umu-C* genotoxicity assays. *Environ. Toxicol. Water Qual.*, 10, pp. 301–305, 1994.

30. Cook, C.R. and S. Chandrasekaran. The start-up and performance of an aerated lagoon and its impact on receiving water quality. *Pulp Paper Can.*, 87, pp. 72–78, 1986.

31. Munro, F.C. Mill experience using oxygen delignification, high substitution, and effluent treatment to reduce dioxin and organochlorine generation and discharge. *Aust. N. Z. Pulp Paper Ind. Tech. Assoc. (APPITA)*, 43, pp. 421–525, 1990.

32. Ontario Ministry of Environment. The Fluctuation Test. Watershed Management Section, Biohazard Unit, 1991.

33. Hubbard, S.A., M.L. Green, D. Gatehouse, and J.W. Bridges. The Fluctuation Test in Bacteria, in *Hand Book of Mutagenicity Test Procedures, 2nd ed.*, Kilby, B.J., M. Legator, W. Nichols, and C. Ramel, Eds., Elsevier Science Publishers, BV, 1984.

34. Gilbert, R.I. The analysis of fluctuation tests. *Mutation Res.*, 74, pp. 283–289, 1980.

35. Miller, J.H. *Assay of ß-Galactosidase. Experiments in Molecular Genetics.* Cold Spring Harbor Laboratory Press, Cold Spring Harbor, New York, pp. 352–355, 1972.

36. Xu, H., B.J. Dutka, and K.K. Kwan. Genotoxicity studies on sediments using a modified SOS chromotest. *Toxicity Assessment*, 2, pp. 79–88, 1987.

37. Dutka, B.J. Methods for Microbiological and Toxicological Analysis of Waters, Waste Waters and Sediments. R.R.B., N.W.R.I., C.C.I.W., Burlington, Ontario, Canada, 1989.

38. Holmbom, B., L. Kronberg, P. Backlund, V-A. Långvik, J. Hemming, M. Reunanen, A. Smeds, and L. Tikkanen. Formation and Properties of 3-Chloro-4-(dichloro-methyl)-5-hydroxy-2(5H)-furanone, a Potent Mutagen in Chlorinated Waters, in *Water Chlorination-Chemistry, Environmental Impact and Health Effects, Vol. 6*, Jolley, R.L., L.W. Condie, J.D. Johnson, S. Katz, R.A. Minear, J.S. Mattice, and V.A. Jacobs, Eds., Lewis Publishers, Boca Raton, FL, 1990.

39. Heddle, J.A., M.C. Cimino, M. Hayashi, F. Romagna, M.D. Shelby, J.D. Tucker, Ph. Vanprays, and J.T. MacGregor. Micronuclei as an index of cytogenetic damage, past, present, and future. *Environ. Mol. Mutagenesis,* 18, pp. 277–291, 1991.

40. Horth, H., M. Fielding, H.A. James, M.J. Thomas, T. Gibson, and P. Wilcox. Production of Organic Chemicals and Mutagens During Chlorination of Amino Acids in Water, in *Water Chlorination-Chemistry, Environmental Impact and Health Effects, Vol. 6*, Jolley, R.L., L.W. Condie, J.D. Johnson, S. Katz, R.A. Minear, J.S. Mattice, and V.A. Jacobs, Eds., Lewis Publishers, Boca Raton, FL, 1990.

41. Holmbom, B. Mutagenic Compounds from Chlorination of Humic Substances, in *Humic Substances in the Aquatic and Terrestrial Environment,* Allard, B., H. Borén, and A. Grimvall, Eds., Springer-Verlag, Berlin, 1991.

Chapter 4

Characteristics of EROD Induction Associated with Exposure to Pulp Mill Effluent

K.R. Munkittrick, M.R. Servos, K. Gorman, B. Blunt, M.E. McMaster, and G.J. Van Der Kraak

INTRODUCTION

Fish collected near pulp mills have consistently shown induction of the hepatic mixed function oxygenase (MFO) detoxification system,[1-14] although the intensity and duration of this induction varies with the species of fish, the sex, the season, and the site examined. These recent North American findings support the earlier European work published to date.[15-19] Although this enzyme induction was originally documented downstream of mills using chlorine with only primary treatment, it has recently been seen at mills which did not use chlorine bleaching[4,20,21] as well as at mills with secondary treatment.[4,5,22] The identity of the inducing chemical(s) is not known.

In addition to increased MFO activity, several other types of changes in fish exposed to effluent from bleached kraft pulp mills have been seen, including delayed sexual maturity, smaller gonads, reduced secondary sexual characteristics, and increased liver size.[1,22] Detailed studies at one site did not find recovery of these responses in wild fish after installation of secondary treatment.[2,23] Expanded studies have shown that most large Ontario mills which were surveyed showed either increased MFOs, decreased steroids, or smaller gonad sizes in female fish.[4] Two of the mills which did not show appreciable steroid impacts during this survey have shown some effects in other years at both Espanola[5] (Servos and Van Der Kraak, unpublished data) and Kapuskasing.[24] Furthermore, the effluent from the mill at Espanola has affected maturity, fecundity, secondary sexual characteristics, and production of steroid hormones in fathead minnow life-cycle studies, at concentrations slightly above those seen in the discharge environment.[25]

The relationship between MFO induction and the reproductive changes is not understood, although there has been speculation that increased MFO activity may lead to increased clearance of endogenous compounds, including reproductive steroids;[26–28] e.g., the elevated MFOs are not directly responsible for the decreased levels of circulating steroid hormones. At Jackfish Bay, circulating steroid hormone levels in exposed fish of various species are consistently 20 to 40% of reference levels at various times of the year,[23] white sucker produce 30% of the 12,000 pg mL^{-1} of circulating levels of testosterone seen in female fish at the reference site at spawning time and still only 30% of the 600 pg mL^{-1} of circulating testosterone during early recrudescence.[23] These observations strongly suggested that the biochemical dysfunction associated with reduced steroid levels was related to a change in a regulatory pathway and not a dysfunction associated with varying levels of an anthropogenic chemical or varying responses to different levels of induction. In fact, steroid depressions have been seen in female fish which did not show induction, and induction has been seen in male fish which did not show depressions in steroid levels.[4]

Detailed studies determined that the reduced circulating levels of steroid hormones were related to a decreased synthesis and not due to an increased catabolic rate.[29] There are several dysfunctions in the control of steroid function, including decreased sensitivity to gonadotropin-releasing hormone analogues, reduced circulating levels of gonadotropin (GtH), decreased sensitivity of ovarian follicles to GtH, altered metabolism of steroids,[29] altered activity of anabolic enzymes in the steroid synthetic pathway, and possibly reduced stores or reduced mobilization of the cholesterol precursor.[30] Although increased clearance rates of steroids cannot account for the differences, it is possible that the reproductive dysfunctions are due to an indirect effect associated with the induction of MFOs, or due to a direct effect of the inducing chemical on receptors regulating the steroid biosynthetic pathway.

We conducted a series of field and laboratory experiments to examine the characteristics of this EROD induction at two pulp mill locations. This included an examination of EROD induction versus distance from the outfall, recovery of EROD induction following mill shutdown, an examination of the source of induction, and a thorough examination of the kinetics of EROD induction.

EXPERIMENTAL METHODS

The study sites are described in detail elsewhere.[1,4,5] Jackfish Bay receives the effluent from a large bleached kraft mill located in Terrace Bay, Ontario; the mill discharged primary-treated effluent until the fall of 1989 when the mill completed installation of an aerated stabilization basin with a 7–10 days retention time. The Terrace Bay mill discharges its effluent into the headwaters of Blackbird Creek, which carries the effluent approximately 15 km to Jackfish Bay on Lake Superior. The mill at Espanola is a modernized mill which has had secondary treatment since the early 1980s, and discharges its effluent into the Spanish River, which eventually reaches Lake Huron. Both mills are described in more detail elsewhere.[31]

The MFO measurements were conducted using the catabolism of 7-ethoxyresorufin (ethoxyresorufin-*o*-deethylase activity; EROD), and the original methods are described in detail elsewhere,[32] measurements were conducted on post-mitochondrial supernatants unless otherwise noted. A detailed interlaboratory comparison of the EROD methodology has been previously published.[33] Chemistry methodologies are described elsewhere.[31,34]

To further examine the induction, water from Blackbird Creek and secondary-treated effluent from the bleach kraft mill in Terrace Bay were collected in May 1993. Water samples were collected in 18-L stainless steel containers, filtered through a 1 μm glass fiber filter, acidified with sulfuric acid to pH 2 and double-extracted by stirring with 400 and 300 mL of dichloromethane for 30 minutes. Sample volume was reduced in a rotoevaporator, transferred into methanol, and evaporated to a final volume under a stream of nitrogen. Filters were Soxhlet-extracted with dichloromethane for 18 hours, and extracts were transferred into methanol. Juvenile rainbow trout were exposed for 4 days to extracts in methanol diluted to the original aqueous volumes. Corresponding controls were prepared using the same solvent volumes prepared in the identical manner.

RESULTS AND DISCUSSION

Induction in Wild and Caged Fish

Initial experiments on the clearance of EROD induction were conducted on white sucker collected at Jackfish Bay in 1989, and then sampled immediately or after 4 days of caging in uncontaminated water. The pulp mill effluent-exposed fish caged for 4 days showed a significant decline in activity,[35] although the effects of capture and caging stress of EROD levels were unknown. However, studies conducted during a mill maintenance shutdown in the fall of 1990 showed rapid recovery of EROD levels in wild longnose sucker, and an apparent recovery of steroid hormones in male white and longnose sucker.[2] Various attempts to reproduce these results over the next two years were unsuccessful due to the extreme variability in EROD induction levels seen at Jackfish Bay; EROD activity varied from 5 to >100 pmol^{-1} min^{-1} mg protein^{-1} (relative to 0.5 to 3.0 at the

Lake Superior reference sites), and survival of the fish after summer gillnetting was poor in caging studies. The wild fish at Jackfish Bay were not susceptible to capture by other methods (trapnetting, electroshocking) during their period of residency in the effluent plume. We decided to attempt caging studies using clean fish to decrease the variability and stress levels on the fish.

Early studies on wild fish showed that the longnose sucker demonstrated a decreased level of EROD activity as distance from the source increased.[2] Several sets of caging studies were conducted using small (<12 cm) white sucker caged for 4 days at various distances from the mouth of the creek which carries the effluent to Jackfish Bay. The cages which were used consisted of two plastic 20-L laundry baskets which were secured together and anchored in place. These results showed a decreasing EROD response with distance from the effluent (Figure 4.1), and decreasing exposure as determined by chlorophenol concentrations (Table 4.1; August data).

Similar studies in the Spanish River showed a rapid induction of EROD activity in caged juvenile white sucker, and the induction persisted 28 km downstream (Figure 4.2). Wild adult white sucker in the Spanish River showed EROD induction up to 51 km downstream which was consistent with a minor reduction in exposure.[5] The sewage treatment plant located less than 2 km downstream of the mill outfall which was potentially a second source of contaminants did not cause an increase in EROD induction in caged fish relative to the site immediately below the outfall (Figure 4.2).

Caging Studies During Mill Shutdowns

During the fall of 1990, we had determined that wild longnose sucker showed a full recovery of EROD activity after a short mill maintenance shutdown.[2] During 1991 and 1992, we attempted to replicate this result using small (<12 cm) white sucker caged at increasing distances from the effluent for 4 days prior to (Figure 4.1, Jackfish Bay), during, and after mill maintenance shutdowns at the pulp mills at Jackfish Bay and on the Spanish River. The induction in white sucker caged in the Spanish River disappeared during two mill maintenance shutdowns in 1992 and reappeared with initiation of mill discharge (Figure 4.3).

Experiments at Jackfish Bay were not as easy to interpret. While the exposures demonstrated a distance-response relationship (Figure 4.1), interpretation of the shutdown data (Figure 4.4a) was complicated by geographic and environmental factors. The transit time for effluent in the creek is >2 days, and the September shutdown period was characterized by several periods of heavy rainfall. Caged fish still showed marked induction during shutdown and showed a clear relationship with distance from the mouth of the Creek (foam barrier). During

August

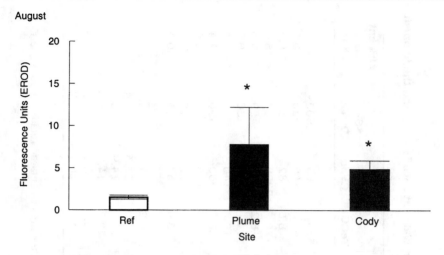

Figure 4.1. EROD activity in white sucker (42 cm) caged in Jackfish Bay for 4 days during August of 1991. Plume is approximately 500 m from the mouth of Blackbird Creek and Cody is approximately 2 km out.

shutdown, the rainfall appeared to flush effluent out of the creek, and fish caged at the mouth of Blackbird Creek were still exposed to >1% effluent (Table 4.1).

To further complicate interpretation, the post-shutdown (October) exposure period was complicated by exceptionally high rainfall (Figure 4.5), and very little induction in fish caged in the effluent was observed (Figure 4.4b); relative exposure levels in Jackfish Bay were lower than during mill shutdown (Table 4.1).

Source of the Induction

Although induction at Jackfish Bay could be clearly associated with exposure to the effluent as it left Blackbird Creek, there was some concern that the long history of sediment deposition in the Creek was contributing to the level of induction. To test this theory, we examined white sucker and rainbow trout exposed to water samples collected at various locations in the Blackbird Creek system. Samples were collected at the discharge point from the secondary treatment system (Effluent), at two crossings where the Creek intersected the highway on its way to Lake Superior (2X and 3X), and at the foam barrier where the creek reaches Lake Superior. The residency time of effluent in the creek system is 2–3 days, and sample collections were staggered to approximate travel times down the creek. Reference water was taken from the unexposed spawning site in Sawmill Creek. White sucker and rainbow trout were exposed in static exposures for 4 days, to 100%, 50%, 25%, and 12.5% of the receiving water samples.

Table 4.1. Levels of Pulp Mill Related Compounds (ng L⁻¹) Measured in Receiving Water at Jackfish Bay and the Spanish River During 1991.[a]

Date	Site	n	246-TCP	2346-TeCP	45-DCG	345-TCG	456-TCG	TeCG	6-CVan	56-DCVan	Effluent Conc. (%)[b]
Jackfish											
Aug.	foam	2	350	250	940	6000	940	570	5700	2800	48
	plume	2	84	43	270	770	150	97	840	320	9
	cody	2	4	2.3	22	39	8.7	8.5	82	30	0.5
Sept.	foam	2	97	5.9	360	650	210	120	2200	510	5
	plume	2	12	<DL	46	88	23	8.2	190	59	0.6
	brook	2	13	5.7	48	86	19	14	140	56	0.6
	cody	2	6.8	1.7	39	48	16	10	140	39	0.4
Oct.	foam	2	170	37	150	740	470	68	1800	490	7.4
	plume	2	12	2.1	18	50	27	9.7	140	42	0.6
	brook	2	<DL	1.2	13	26	14	5.6	69	18	0.4
	cody	2	4.4	1.3	11	18	9.2	3.9	62	15	0.3
Espanola											
July 11	28 km	2	44.7	2.1	37	25.9	11.6	14.2	144	3.9	
Aug. 4	28 km	2	6.3	1.0	23.2	17.6	3.5	6.0	33.3	7.4	
Sept. 3	8 km	2	62.5	2.9	15.7	33.3	22.9	28.9	14.9	6.1	
Sept. 8	28 km	2	172.6	3.9	28.9	22.6	12.5	20.7	14.9	5.5	
Sept. 14	28 km	2	54.7	1.9	18.2	9.8	6.4	8.7	8.2	4.0	
Oct. 7	28 km	2	<DL	<DL	27.2	17.9	<DL	13.7	20.4	13.2	

[a] Levels at the reference sites were non-detectable for all sampling times.
[b] Foam barrier estimates based on approximate ratios of chlorophenols in effluent. Other sites calculated relative to foam barrier.

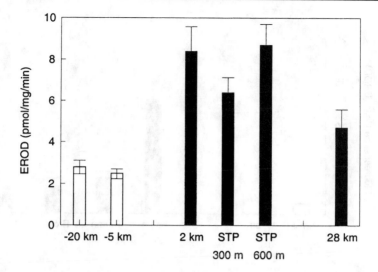

Figure 4.2. EROD activity in white sucker (<10 g body weight) caged at 2 sites upstream of the discharge from a bleached kraft pulp and paper mill, and at sites 2 km downstream, adjacent to a sewage treatment plant, and 28 km downstream of the discharge.

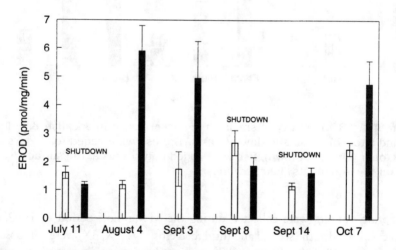

Figure 4.3. EROD activity in small white sucker caged 28 km downstream of the discharge in the Spanish River (shaded box) during mill maintenance shutdowns, and during normal discharge periods, relative to caging at an upstream reference site.

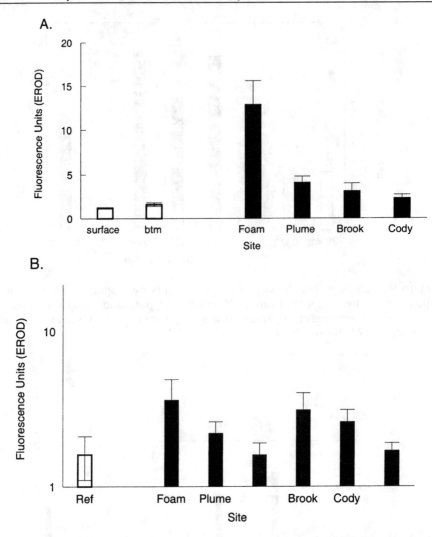

Figure 4.4. EROD activity in small white sucker caged in Jackfish Bay during (A) and after (B) a mill shutdown (1991). The fish were caged for 4 days, and in October (after shutdown), fish were caged at two sites at the surface (solid), and on the sediments (hatched lines).

Spring samples were collected during the May runoff period in 1992, and samples collected at the second and third road crossings (2X, 3X) were composed of approximately 40–60% effluent, while the foam barrier was composed of approximately 30% effluent (Table 4.2). These values are consistent with the contributions of additional streamflows to the Blackbird Creek system. During May, both rainbow trout and white sucker showed significant induction in only 100 and 50%

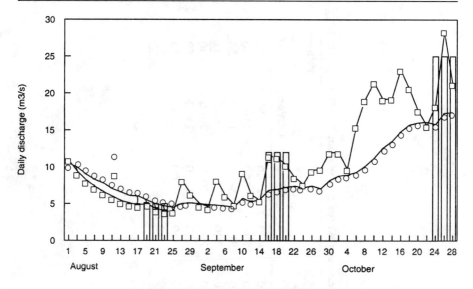

Figure 4.5. River discharge data (Water Survey of Canada) for gauges located on two rivers (Steele (circles, gauge 02BA002), Little Pic (boxes, gauge 02BA003)) near Blackbird Creek, Lake Superior. The bars represent periods during which fish were caged in Jackfish Bay.

effluent, and in the 100% exposure of the second crossing (Figure 4.6a,b). There was no evidence of the receiving system contributing to the level of induction.

Fall samples were collected during a relatively low flow period during September 1992, which resulted in higher percent effluent concentrations in the receiving areas. The EROD data from rainbow trout also shows much less loss of potency with distance downstream (Figure 4.6c), and again there was no evidence of an increased induction after effluent transport in the Creek system. In both May (white sucker) and September (rainbow trout) exposures, fish exposed to 25% secondary effluent showed higher induction than fish exposed to 100% of receiving water from any of the downstream sites.

Extraction of the Inducers

Exposure of rainbow trout to whole water from Blackbird Creek caused rapid induction within 4 days and induction after 8 days was similar (data not shown). Extracts of the water or filters from Blackbird Creek or the reference site (Mountain Bay) reconstituted to the original volume did not cause induction above the levels observed in laboratory reference water exposures (Figure 4.7a). However, increasing the extract concentration by four times resulted in elevated EROD activity above the controls (Figure 4.7b). Increasing the exposure to the extract

Table 4.2. Levels of Pulp Mill Related Compounds (ng L^{-1}) Measured in Receiving Water at Jackfish Bay During 1992.[a]

Date	Site	n	246-TCP	2346-TeCP	45-DCG	345-TCG	456-TCG	TeCG	6-CVan	56-DCVan	Effluent Conc. (%)
May	effluent	5	220	<DL	1400	13000	5400	1900	2200	640	100
	2X	4	120	<DL	860	4800	2100	790	5500	2100	50
	3X	4	170	<DL	980	5600	2300	890	3800	1400	56
	foam	4	78	<DL	560	3000	1200	620	1700	620	31
Sept.	effluent	2	2800	<DL	2700	4600	3700	260	7000	2000	100
	2X	2	310	37	2500	3000	1800	95	4700	1400	69
	3X	3	270	42	2300	3300	2100	320	4200	1200	73
	foam	2	540	73	650	1100	790	48	2600	910	22

[a] Levels at the reference site were below detection limits during both sampling periods.

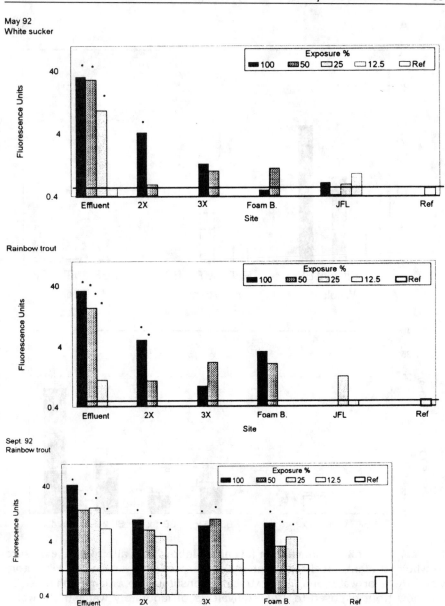

Figure 4.6. Levels of hepatic EROD in white sucker or rainbow trout after exposure to diluted concentrations of Terrace Bay effluent or receiving water collected at increasing distances downstream in Blackbird Creek. The sites at 2X and 3X represent road crossings of Blackbird Creek, the foam barrier is at the mouth of the creek at Lake Superior, and JFL represents control caging at Jackfish Lake (unexposed to effluent). The solid line represents two standard errors above the mean activity of reference fish.

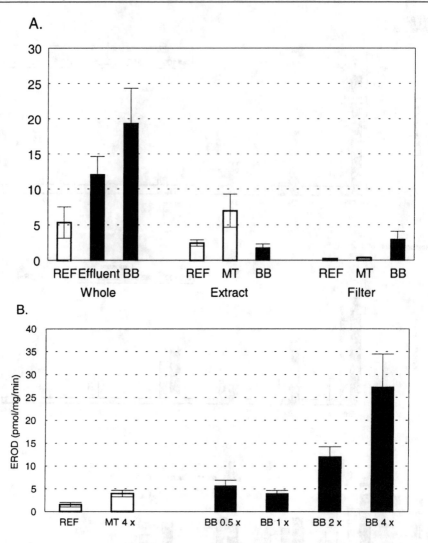

Figure 4.7. (a) EROD induction in juvenile rainbow trout after 4 days exposure to whole effluent, receiving water (BB), reference site water (MT), laboratory water (Ref), or water and filter extracts reconstituted to original volumes. (b) Rainbow trout exposed to 12-L equivalents of laboratory water (Ref), water from the reference site (MT), and to 0.5, 1, 2 and 4 x (12 L equivalents) of Blackbird Creek water.

caused induction suggesting poor extraction efficiency or partial degradation during extraction may have reduced the concentration of the inducing compounds in the extracts. The results demonstrate that a simple solvent extraction can isolate and concentrate the inducing chemicals.

Experiments have shown that induction is not affected by secondary treatment[2] or centrifugation and filtering.[36] Detailed experiments have shown that the induction is not caused by any of the dominant resin acids found in effluent.[37] Hewitt et al.[36] compared EROD induction with chemical levels in effluent after primary and secondary treatment, as well as after filtration. The conclusions of those experiments were that levels of resin acids, fatty acids, terpenes, chlorophenolics, aliphatic alkanes, plant sterols, and chlorinated dimethyl-sulphones did not correlate with EROD induction potential. Various research groups are attempting to identify the chemical(s) responsible for the induction associated with pulp mill effluent.[38]

Kinetics of Induction

In an attempt to control the exposures better, in May of 1992 white sucker were placed into 80-L buckets with Blackbird Creek water collected from the third crossing (3X) and was exchanged every 2 days. The effluent concentration in the Creek water was approximately 50% (Table 4.2). Small (<2 g) white sucker were placed in the effluent for 2, 4, 8, or 14 days, and then allowed to depurate for 2, 4, or 8 days in reference water.

These results showed that induction was very rapid and occurred within 2 days of exposure (Figure 4.8). If fish were removed to clean water after 4 days of exposure, activity continued to increase during the depuration phase, suggesting that the inducers were not being excreted from the body. However, when fish were removed to clean water after 14 days of exposure, the EROD activity rapidly declined to baseline levels (Figure 4.8). These experiments concluded that the inducer(s) appeared to be excreted by an inducible mechanism, that it was rapidly metabolized once the mechanism was induced, and that the inducing compound(s) were rapidly excreted and unlikely to biomagnify or accumulate in the bodies of the fish to appreciable levels. These experimental results have predominately used secondary-treated effluent from two relatively modern, bleached kraft mills. It is possible that induction at some other sites, and in some other species of fish, may be associated with different compounds or have different characteristics. It is clear from the Jackfish Bay studies that the fish are capable of excreting the inducing compound(s) inherent in the Jackfish Bay effluent, but this observation cannot be extended to all effluents.

In an attempt to determine whether EROD induction was the mechanism responsible for the excretion of the inducers, rainbow trout were induced with 3 mg kg^{-1} of β-napthoflavone (BNF) 3 days prior to exposure to pulp mill effluent. These experiments were conducted in the laboratory, and preliminary trials demonstrated that the induction associated with BNF would be reduced to reference levels in 4–8 days. The BNF-exposed fish showed a classic decline in activity after

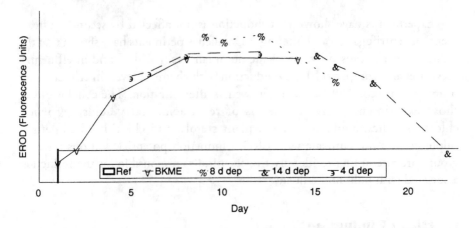

Figure 4.8. Levels of hepatic EROD in small (<2 g) white sucker exposed to Blackbird Creek water (third crossing, approx. 50% effluent) for 2, 4, 8, or 14 days. Exposures were conducted in 80-L plastic buckets, with water exchanged every 2 days. EROD estimates were conducted on 500 µL of crude liver homogenates.

initiation of the exposure trials (Figure 4.9a). Fish preexposed to BNF prior to effluent exposure showed a lower EROD activity after 16 days in effluent (Figure 4.9a), although there were no apparent differences in recovery profiles relative to effluent-exposed fish after 2 or 4 days of exposure (Figures 4.9b,c). It is unclear whether preexposure to BNF changed the response of fish to the effluent.

CONCLUSIONS

Although the causative agents for the EROD induction associated with pulp mills are not known, several characteristics of the inducer(s) are apparent from these and other studies:

- EROD induction has been found after exposure to effluent from bleached kraft, sulfite and thermo-mechanical pulp mills, and is not affected by removal of chlorine use or installation of secondary treatment.
- The inducer is persistent in the water column for more than 50 km downstream of mills, and the threshold for induction is <1% effluent at some mill sites.
- The inducer disappears from the water column very quickly during mill maintenance shutdowns.
- In field trials, uptake of the inducer is very rapid and EROD induction peaks within 2 days in warm water.

A.

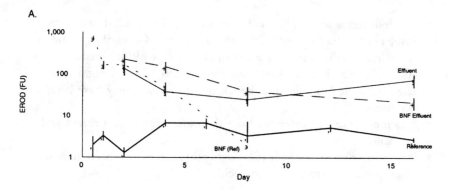

B.

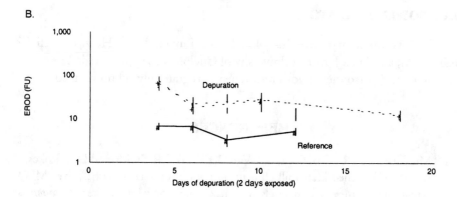

C.

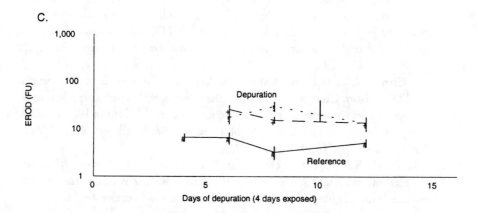

Figure 4.9. (a) Levels of hepatic EROD in rainbow trout exposed to effluent from the Terrace Bay mill, with or without preinduction with β-napthoflavone (BNF). Fish were injected 3 days prior to initiation of effluent exposures. (b) Levels of EROD in rainbow trout after 2 days exposure and 2, 4, 8, or 16 days in clean water. (c) Levels of EROD in rainbow trout after 4 days exposure and 2, 4, or 8 days in clean water.

- The inducers at some mills can be excreted rapidly by fish, but excretion involves an inducible mechanism which requires >4 days of exposure; it is unclear whether the mechanism of excretion is EROD mediated.
- At Jackfish Bay it seems likely that the inducer(s) is (are) not accumulated from the food chain, although this historically may have been a component.
- The inducer can be extracted from the water column and eluted with methanol.
- The inducer(s) are not likely to be highly chlorinated dioxins, resin acids, chlorophenolics, sterols, or fatty acids.

ACKNOWLEDGMENTS

The assistance of Beth Chisholm, Lynne Luxon, Mark Hewitt, Weli Xi (GLLFAS), and Janet Jardine (University of Guelph) was appreciated. The cooperation of mill personnel at both research sites is gratefully acknowledged.

REFERENCES

1. Munkittrick, K.R., C.B. Portt, G.J. Van Der Kraak, I.R. Smith, and D.A. Rokosh. Impact of bleached kraft mill effluent on population characteristics, liver MFO activity and serum steroid levels of a Lake Superior white sucker (*Catostomus commersoni*) population. *Can. J. Fish. Aquat. Sci.*, 48, pp. 1371–1380, 1991.
2. Munkittrick, K.R., G.J. Van Der Kraak, M.E. McMaster, and C.B. Portt. Response of hepatic mixed function oxygenase (MFO) activity and plasma sex steroids to secondary treatment and mill shutdown. *Environ. Toxicol. Chem.*, 11, pp. 1427–1439, 1992.
3. Munkittrick, K.R., M.E. McMaster, C.B. Portt, G.J. Van Der Kraak, I.R. Smith, and D.G. Dixon. Changes in maturity, plasma sex steroid levels, hepatic mixed-function oxygenase activity, and the presence of external lesions in lake whitefish (*Coregonus clupeaformis*) exposed to bleached kraft mill effluent. *Can. J. Fish. Aquat. Sci.*, 49, pp. 1560–1569, 1992.
4. Munkittrick, K.R., G.J. Van Der Kraak, M.E. McMaster, C.B. Portt, M.R. van den Heuvel, and M.R. Servos. Survey of receiving-water environmental impacts associated with discharges from pulp mills. 2. Gonad size, liver size, hepatic EROD activity and plasma sex steroid levels in white sucker. *Environ. Toxicol. Chem.*, 13, pp. 1089–1101, 1994.
5. Servos, M.R., J.H. Carey, M.L. Ferguson, G.J. Van Der Kraak, H. Ferguson, J. Parrott, K. Gorman, and R. Cowling. Impact of a modernized bleached kraft mill on white sucker populations in the Spanish River, Ontario. *Wat. Pollut. Res. J. Can.*, 27, pp. 423–437, 1992.

6. Hodson, P.V., M. McWhirter, K. Ralph, B. Gray, D. Thivierge, J.H. Carey, G. Van Der Kraak, D.M. Whittle, and M.C. Levesque. Effects of bleached kraft mill effluent on fish in the St. Maurice River, Quebec. *Environ. Toxicol. Chem.,* 11, pp. 1635–1651, 1992.

7. Adams, M.S., W.D. Crumby, M.S. Greeley, Jr., L.R. Shugart, and C.F. Saylor. Responses of fish populations and communities to pulp mill effluents—A holistic assessment. *Ecotoxicol. Environ. Saf.,* 24, pp. 347–360, 1992.

8. Mather-Mihaich, E. and R.T. Di Guiglio. Oxidant, mixed function oxidase and peroxisomal responses in channel catfish exposed to a bleached kraft mill effluent. *Arch. Envion. Contam. Toxicol.,* 20, pp. 391–397, 1991.

9. Smith, I.R., C.B. Portt, and D.A. Rokosh. Hepatic mixed function oxidases induced in populations of white sucker, *Catostomus commersoni,* from areas of Lake Superior and the St. Mary's River. *J. Great Lakes Res.,* 17, pp. 382–393, 1991.

10. Smith, I.R., C.B. Portt, and D.A. Rokosh. The impact of two Great Lakes bleached kraft mills (BKME) on receiving water mutagenicity and the biochemistry and pathology of wild white suckers inhabiting impacted areas. *Can. Tech. Rep. Fish. Aquat. Sci.,* 2, pp. 718–719, 1991.

11. Rogers, I.H., C.D. Levings, W.L. Lockhart, and R.J. Norstrom. Observations on overwintering juvenile chinook salmon (*Oncorhynchus tshawytscha*) exposed to bleached kraft mill effluent in the upper Fraser River, British Columbia. *Chemosphere,* 19, pp. 1853–1868, 1989.

12. Boyle, D.E., B.A. Bravender, T.J. Brown, D. Kieser, C.D. Levings, W.L. Lockhart, J.A. Servizi, and T.R. Whitehouse. Baseline monitoring of mountain whitefish, *Prosopium williamsoni,* from the Columbia River system near Castlegar, British Columbia: Health, contaminants and biology. *Can. Tech. Rep. Fish. Aquat. Sci.,* 1883, 1992.

13. Ahokas, J.T., D.A. Holdway, S.E. Brennan, R.W. Goudey, and H.B. Bibrowska. MFO activity in carp (*Cyprinus carpio*) exposed to treated pulp and paper mill effluent in Lake Coleman, Victoria, Australia, in relation to AOX, EOX, and muscle PCDD PCDF. *Environ. Toxicol. Chem.,* 13, pp. 41–50, 1994.

14. Courtenay, S., C. Grunwald, G.L. Kreamer, R. Alexander, and I. Wirgin. Induction and clearance of cytochrome P4501A messenger-RNA in Atlantic tomcod caged in bleached kraft mill effluent in the Miramichi River. *Aquat. Toxicol.,* 27, pp. 225–243, 1993.

15. Andersson, T., B.-E. Bengtsson, L. Förlin, J. Härdig, and Å. Larsson. Long term effects of bleached kraft mill effluents on carbohydrate metabolism and hepatic xenobiotic biotransformation enzymes in fish. *Ecotoxicol. Environ. Saf.,* 13, pp. 53–60, 1987.

16. Andersson, T., L. Förlin, J. Härdig, and Å. Larsson. Physiological disturbances in fish living in coastal water polluted with bleached kraft pulp mill effluents. *Can. J. Fish. Aquat. Sci.,* 45, pp. 1525–1536, 1988.

17. Lindström-Seppä, P. and A. Oikari. Biotransformation and other physiological responses in whitefish caged in a lake receiving pulp and paper mill effluents. *Ecotoxicol. Environ. Saf.,* 18, pp. 191–203, 1989.

18. Lindström-Seppä, P. and A. Oikari. Biotransformation activities of feral fish in waters receiving bleached pulp mill effluents. *Environ. Toxicol. Chem.,* 9, pp. 1415–1424, 1990.

19. Lindström-Seppä, P. and A. Oikari. Biotransformation and other toxicological and physiological responses of rainbow trout (*Salmo gairdneri richardson*) caged in a lake receiving effluents of pulp and paper industry. *Aquat. Toxicol.,* 16, pp. 187–204, 1990.

20. Gagne, F. and C. Blaise. Hepatic metallothionein level and mixed function oxidase activity in fingerling rainbow trout *(Oncorhynchus mykiss)* after acute exposure to pulp and paper mill effluents. *Wat. Res.,* 27, pp. 1669–1682, 1993.

21. Martel, P.H., T.G. Kovacs, B.I. O'Connor, and R.H. Voss. A survey of pulp and paper mill effluents for their potential to induced mixed function oxidase enzyme activity in fish. *Wat. Res.,* 28, pp. 1835–1844, 1994.

22. McMaster, M.E., G.J. Van Der Kraak, C.B. Portt, K.R. Munkittrick, P.K. Sibley, I.R. Smith, and D.G. Dixon. Changes in hepatic mixed function oxygenase (MFO) activity, plasma steroid levels and age at maturity of a white sucker (*Catostomus commersoni*) population exposed to bleached kraft pulp mill effluent. *Aquat. Toxicol.,* 21, pp. 199–218, 1991.

23. Munkittrick, K.R., G.J. Van Der Kraak, M.E. McMaster, and C.B. Portt. Reproductive dysfunction and MFO activity in three species of fish exposed to bleached kraft mill effluent at Jackfish Bay, Lake Superior. *Wat. Poll. Res. J. Can.,* 27, pp. 439–446, 1992.

24. Nickle, J.C., M.E. McMaster. K.R. Munkittrick, C. Rumsey, C. Portt, and G.J. Van Der Kraak. Reproductive Effects of Primary-Treated Bleached Kraft and Thermomechanical Pulp Mill Effluents on White Sucker in the Moose River Basin. 3rd International Conference on Environmental Fate and Effects of Pulp and Paper Mill Effluents, Rotorua, NZ, November 9–13, 1997.

25. Robinson, R.D. Evaluation and development of laboratory protocols for predicting the chronic toxicity of pulpmill effluents to fish. Ph.D. Thesis, University of Guelph, 1994.

26. Peakall, D.B. *Animal Biomarkers as Pollution Indicators.* Chapman & Hall, Ecotoxicology Series, Vol. 1, Chapman & Hall, New York, 1992.

27. Lee, R.F. Possible Linkages Between Mixed Function Oxygenase Systems, Steroid Metabolism, Reproduction, Molting, and Pollution in Aquatic Animals, in *Toxic Contaminants and Ecosystem Health: A Great Lakes Focus,* Evans, M.S., Ed., John Wiley & Sons, New York, 1988.

28. Okey, A. Enzyme induction in the cytochrome P-450 system. *Pharmacol. Therapeut.,* 45, pp. 241–298, 1990.

29. Van Der Kraak, G.J., K.R. Munkittrick, M.E. McMaster, C.B. Portt, and J.P. Chang. Exposure to bleached kraft pulp mill effluent disrupts the pituitary

gonadal axis of white sucker at multiple sites. *Toxicol. Appl. Pharmacol.*, 115, pp. 224–233, 1992.

30. McMaster, M.E., G.J. Van Der Kraak, and K.R. Munkittrick. Evaluation of the steroid biosynthetic capacity of ovarian follicles from white sucker exposed to bleached kraft mill effluent (BKME). *Can. Tech. Rep. Fish. Aquat. Sci.*, p. 259, 1993.

31. Robinson, R.D., J.H. Carey, K.R. Solomon, I.R. Smith, M.R. Servos, and K.R. Munkittrick. Survey of receiving water environmental impacts associated with discharges from pulp mills. 1. Mill characteristics, receiving water chemical profiles and laboratory toxicity tests. *Environ. Toxicol. Chem.*, 13, pp. 1075–1088, 1994.

32. McMaster, M.E., C.B. Portt, K.R. Munkittrick, and D.G. Dixon. Milt characteristics, reproductive performance, and larval survival and development of white sucker exposed to bleached kraft mill effluent. *Ecotox. Environ. Saf.*, 23, pp. 103–117, 1992.

33. Munkittrick, K.R., M.R. van den Heuvel, J.J. Stageman, W.L. Lockhart, and D.A. Metuer. Interlaboratory comparison and optimization of the ethoxyresorufin-*o*-deethylase determination for hepatic monooxygenase activity in white sucker (*Catostomus commersoni*) exposed to bleached kraft pulp mill effluent. *Environ. Toxicol. Chem.*, 12, 1273–1282, 1993.

34. Servos, M.R., S.Y. Huestis, D.M. Whittle, G.J. Van Der Kraak, and K.R. Munkittrick. Survey of receiving water environmental impacts associated with discharges from pulp mills. 3. Polychlorinated dioxins and furans in muscle and liver of white sucker (*Catostomus commersoni*). *Environ. Toxicol. Chem.*, 13(7), pp. 1103–1115, 1994.

35. McMaster, M.E. Bleached Kraft Pulp Mill Effluent and Its Impacts on Fish Populations in Jackfish Bay, Lake Superior, M.Sc. Thesis, Dept. of Biology, University of Waterloo, Waterloo, Ontario, Canada, 1991.

36. Hewitt, L.M., J.H. Carey, K.R. Munkittrick, and D.G. Dixon. Examination of Bleached Kraft Mill Effluent Fractions for Potential Inducers of Mixed Function Oxygenase Activity in Rainbow Trout, in *Environmental Fate and Effects of Pulp and Paper Mill Effluents*, Servos, M.R., K.R. Munkittrick, J.H. Carey, and G. Van Der Kraak, Eds., St. Lucie Press, DelRay Beach, FL, 1996, pp. 79–94.

37. Ferguson, M.L., M.R. Servos, K.R. Munkittrick, and J. Parrott. Inability of resin acid exposure to elevate EROD activity in rainbow trout (*Oncorhynchus mykiss*). *Wat. Poll. Res. J. Can.*, 27, pp. 561–574, 1992.

38. Martel, P.H., T.G. Kovacs, B.I. O'Connor, and R.H. Voss. The source and identity of compounds in a thermomechanical pulp mill effluent inducing hepatic mixed function oxygenase (MFO) activity in fish. *Environ. Toxicol. Chem.*, 16, 2375–2383, 1997.

Chapter 5

EROD Induction in Fish: A Tool to Measure Environmental Exposure

J.L. Parrott, R. Chong-Kit, and D.A. Rokosh

INTRODUCTION

Mixed function oxygenases (MFOs) are a family of inducible enzymes which oxidize, by single oxygen addition, natural and anthropogenic chemicals. Their metabolic function assists in the excretion of nonpolar compounds. These enzymes may be isolated with the microsomal fraction of cell extracts. The operation of these enzymes requires oxygen and a reducing agent, such as reduced NADP, with electron transport facilitated through cytochrome P450.[1]

Inducible MFO enzymes increase in activity, from constitutive to maximum levels, as a function of dose of an inducing agent such as 2,3,7,8-tetrachloro-dibenzo-*p*-dioxin (2378-TCDD).[2-4] The increase in MFO activity usually indicates an increase in the amount of enzyme and is referred to as induction.[5]

A number of chemical classes are known inducing agents. In chick and in mammalian systems, polycyclic aromatic hydrocarbons (PAHs) and chlorinated congeners of dibenzo-*p*-dioxins (PCDDs), dibenzofurans (PCDFs), azobenzenes, azoxybenzenes, biphenyls (PCBs), and naphthalenes are reported inducing agents.[5,6] Potency of induction is dependent on the stearic configuration of the compound as well as the number and position of halogen (chlorine or bromine) atoms. The most potent inducers are lipid-soluble, planar compounds of 3 Å × 10 Å size. Lateral halogen substitution confers favorable electronic configuration and higher induction potency. PAHs, which are not halogenated, may also be potent inducers.

Usually, MFO enzymes are quantified by catalytic assays, that measure the reaction rate or activity of enzyme present. Cell extracts are incubated in a reaction mixture containing one of several possible substrates that can be metabolized by MFO enzymes. At time intervals or after a predetermined time, quantities of product generated or substrate lost are measured. Enzyme activity is reported in terms of the rate of reaction per minute and standardized on the amount of protein in the cell extract. Most recognized MFO enzyme assays include: Aryl hydrocarbon hydroxylase (benzo(a)pyrene 3-hydroxy-hydroxylase [AHH]), ethoxyresorufin-O-deethylase (EROD), or 7-ethoxycoumarin-O-demethylase (ECOD). Historically, assays for AHH and EROD have been favored, principally because of the simplicity of their assay system.

There are many different types of MFO enzymes and P450-catalyzed reactions (reviewed by Nebert and Nelson[7]). For the purposes of this overview, we will limit our discussions to enzymes associated with cytochrome P4501A1 with activity of one of the associated enzymes measured as EROD.

This chapter overviews MFO bioassays as tools in the monitoring of environmental toxicology. Methods for the EROD bioassays are described and confounding variables affecting the application of MFO bioassays are considered. Induction by various industrial effluents is compared in field studies, in caging studies and in lab exposures. The meaning of MFO induction, a much-debated topic, is also discussed.

MFO Enzymes in Animal Systems

Although original studies were conducted in chick embryo[8] and mammalian systems,[9,10] MFO enzymes are found in a wide range of animal species. Inducible MFO activity has been measured in fresh water fish[11-13] and marine fish.[14-20] P-450 mediated MFO activity was present in the insect *Drosophila melanogaster*[21] as well as in species of marine crab and polychaete.[22] Inferring from an animal's ability to develop tumors when exposed to promutagenic nitrosamines, monooxygenase systems likely exist in the amphibians *Triturus helveticus,*[23] *Xenopus borealis,*[24] and *Rana temporaria*[25] as well as the crayfish *Procambarus clarkii*[26] and the bivalve mollusk *Unio pictorium.*[27]

MFO in Wild Fish

For environmental toxicologists, induction of MFO is a sensitive response that measures exposure of fish to foreign chemicals. Increased MFO activity is frequently observed in fish captured from waters contaminated by pulp mill effluents[12,13,28,29] or petroleum hydrocarbons.[15,17-20] In the wild, MFO levels in fish

Table 5.1. Fish Species in Which MFO Induction Has Been Demonstrated.

	Reference Number
Freshwater Fish	
Bluegill Sunfish *Lepomis macrochirus*	31
Carp *Cyprinus carpio*	32
Chinook salmon *Oncorhynchus tshawytscha*	28
Rainbow trout *Oncorhynchus mykiss*	33
White sucker *Catostomus commersoni*	12
Mountain whitefish *Prosopium williamsoni*	30
Longnose sucker *Catostomus catostomus*	30
Lake whitefish *Coregonus clupeaformis*	34
Lake trout *Salvelinus namaycush*	34
Catfish *Ictalurus punctatus*	35
Largemouth bass *Micropterus salmoides*	35
Marine Fish	
Blenny *Blennius pavo*	in 36
California flatfish *Citharicthys sordidus*	18
California flatfish *Citharicthys stigmaeus*	18
Cunner *Tautogolabrus adspersus* (Walbaum)	19
Dogfish shark *Squalus acanthias*	37
Flounder *Pseudopleuronectes americanus*	in 36
Little skate *Raja erinacea*	37
Mugil cephalus	32
Pomadasys corvinaeformis	This chapter
Scup *Stenotomus chrysops*	38
Sheepshead minnow *Archosargus probatocephalus*	37
Southern flounder *Paralichthyes lethostigma*	37
Stingray *Dasyatis sabina*	37
Ocean perch *Perca fluviatilis*	39
Killifish *Fundulus heteroclitus*	16
English sole *Parophrys vetulus*	40
Starry flounder *Platichthys stellatus*	40
Dab *Limanda limanda*	20
Sea bass *Dicentrarchus labrax*	41

are usually highest near to the contaminant or effluent source, and decreased at sites further from source.

EROD or AHH activity has been measured in a wide variety of freshwater and marine fish species (Table 5.1). However, several species are insensitive, and show little or no MFO induction. Goldfish (*Carassius auratus*), cyprinids (e.g., fathead minnow *Pimephales promelas*), and burbot (*Lota lota*)[30] appear relatively insensitive, and we found that Peamouth chub (*Mylocheilus caurinus* [Richardson]) showed no induction after exposure to 10 µg β-naphthoflavone (BNF)/L.

Fish Exposure

Field Capture

Wild fish can be captured in their natural pristine or contaminated environments. Enzyme levels, with variance, in affected fish populations are determined by grouping MFO activities in tissues, usually the liver, of individual fish captured in the study area. In North American freshwater environments, the common white sucker (*Catostomus commersoni*) has been extensively used as a test species. Surveys are designed to determine MFO levels at a control location and in waters impacted by a source of contaminants. Such studies are generally restricted to surveys at locations and at times when fish capture is convenient, such as during a spawning run. When populations are abundant, surveys may be extended to include resident and forage habitats.

Fish Caging Exposures

Captured wild fish from a reference location can be transported to affected sites and contained in wire mesh cages, without feeding, for periods of three to four days. MFO activity is measured in fish caged at reference locations and in impacted waters at varying distances from an effluent source. Alternatively, cultured laboratory fish such as rainbow trout can be used in caging studies. Care must be taken to avoid acutely toxic effects due to drastic temperature changes, low dissolved oxygen levels, extreme pH, or lethal contaminant levels. In general, bioassays for MFO induction in caged fish are sufficiently robust to be applied in a range of aquatic environments.

Laboratory Fish Exposures

Cultured laboratory fish are exposed in simple setup aquaria or buckets, with solutions renewed continuously or daily. Rainbow trout surviving a 96-hour acute lethality test may also be tested for MFO induction. Fish are exposed to a dilution series of the test sample or chemical prepared using a reference water diluent, such as dechlorinated city tap water. The dilution series includes a diluent control. Generally 10 fish/dilution are exposed. Livers from individual fish or pooled livers from 10 fish are tested for MFO activity. The standard exposure time is two to four days. Feeding is not necessary for short exposures. Temperature is controlled by water bath or environmental chamber, but control may not be required, depending on species and ambient air temperatures (time of year). Mortality as well as pH, conductivity, and dissolved oxygen are recorded. Care must be taken to maintain an acceptable loading density for fish in aquaria during the course of the exposure period.

MFO ASSAY

In the general assay, the animal is sacrificed and a target organ, usually the liver, is obtained. The organ is physically disrupted and the desired enzyme fraction concentrated by centrifugation in a cell-free homogenate (S9 fraction) or a microsomal pellet. The protein content of the enzyme fraction is measured. MFO activity is determined in an aerobic reaction mixture containing the enzyme fraction, cofactors, an energy source (NADPH), and a suitable substrate (for example, 7-ethoxyresorufin). The product, usually a fluorescent molecule (for example, resorufin), is measured continuously or at the end of the reaction. Disappearance of the substrate can also be measured, as in the AHH assay. With most assays, the amount product is quantified by comparing the measured response (change in fluorescence) with a standard curve. Results are expressed as amount of product per minute per mg of S9 or microsomal protein. MFO specific activity (Sp. Act.) is expressed as pmoles product/min/mg protein. Induction can be expressed as "fold induction" by comparing MFO Sp. Act. in exposed animals to MFO Sp. Act. in concurrent nonexposed control animals.

Ethoxyresorufin-O-Deethylase (EROD) Assay

Preparation of Liver Homogenates or Microsomes

Only viable fish are used to provide tissue for enzyme assays. Fish are sacrificed by a blow to the head or cervical dislocation. Fish livers are rapidly removed and transferred to chilled-on-ice Buffer A consisting of 0.1 M potassium phosphate buffer pH 7.4 supplemented with EDTA (1 mM), dithiothreitol (1 mM), and glycerol (20% v/v). If necessary, livers are washed in isotonic KCl to remove bile. We recommend recording fish length, weight, and liver weight as well as the appearance of the liver during autopsy. Livers are stored cryogenically in liquid nitrogen or in a freezer at $-80°C$.

Livers are accurately weighed (0.01 g to 1.0 g—depending on analysis method), suspended in Buffer A or HEPES-KCl,[42] and homogenized by motorized Potter Ehrehjem apparatus. Small livers are homogenized using a handheld grinder and an Eppendorf centrifuge tube. Livers and homogenates are held on ice. The liver homogenate is centrifuged at $9,000 \times g$ for 20 minutes to provide a supernatant ("S9 fraction") containing the cytosol, protein, blood cells, and microsomes. Further centrifugation of the S9 supernatant at $100,000 \times g$ for one hour produces a pellet containing microsomes. This pellet is washed and resuspended in 0.5 mL buffer to provide a "microsomal fraction." Analysis can be performed on the S9 or microsomal fraction.

Measurement of EROD Activity

The reaction mixture containing 125 μmoles HEPES buffer pH 7.8, 25 μmoles $MgSO_4$, 0.7 μmoles NADPH, and 1 mg bovine serum albumin in a total final volume of 1.43 mL, is brought to a temperature of 25°C. A volume of 0.05 mL microsomal suspension in buffer is added. The reaction is started by the addition of 2.5 nmoles 7-ethoxyresorufin (7-ER) in dimethylsulphoxide (DMSO) spectrophotometrically adjusted to a concentration of 20 μg/mL which corresponded to an Absorbance of 1.75 at 462 nm. The reaction is terminated, after a 15 minute incubation, by the addition of 3.2 mL methanol.

A typical assay would contain duplicate reaction tubes and a reagent blank. The reagent blank contained all the constituents of the reaction tube but methanol was added to the tube prior to addition of 7-ER. Tubes were centrifuged at ~2,500 × g for 15 minutes to precipitate protein. Fluorescence was measured in a Turner model 450 Fluorometer, with narrow-band 520 nm excitation filter and >585 nm emission filter. Fluorescence was compared to a resorufin standard curve with slope of 0.2065 fluorescence units/pmole resorufin at a fluorometer gain setting of 50.

Alternatively, EROD activity was measured in a 96-well plate kinetic assay that followed the reduction of 7-ethoxyresorufin to resorufin over 12 minutes, using a multiwell plate reading fluorometer (Cytofluor 2300, Millipore Ltd.; 530 nm excitation filter, 590 nm emission filter; sensitivity 3).[42]

Protein Determination

Protein was determined by the Bradford method.[43] The S9 supernatant or microsomal suspension (50 μL) was diluted to 1.00 mL in water. Diluted solution (50 μL) was combined with 5 mL of 4:1 diluted Bio-Rad® reagent (Hercules, California). Color was measured spectrophotometrically at 585 nm. Protein concentration was determined by comparison with a standard curve for bovine serum albumin with slope of 0.011 Absorbance Units/mg protein.

Positive and Negative Controls

To ensure that test fish were responsive to inducers, positive control fish were exposed to 10 μg/L β-naphthoflavone (BNF), a known inducer of EROD. Negative control fish were exposed to dilution water or dilution water containing the carrier solvent methanol to determine the natural or constitutive levels of EROD in unexposed fish. To determine repeatability of the enzyme assay, EROD activity is periodically measured in a cryogenically preserved BNF-induced liver preparation (S9 or microsomal fraction) with a known enzyme specific activity.

MFO AS AN ENVIRONMENTAL MONITORING TOOL

MFO Induction in Wild Fish

Many species of fish have been shown to have inducible MFO activities. Induction is a common finding in fish exposed to industrial effluents containing complex mixtures of anthropogenic and natural chemicals. White sucker (*Catostomus commersoni*) captured from a freshwater environment impacted by bleached kraft mill effluent showed 20–40 fold EROD induction relative to fish from a reference site.[44] Larsson et al.[39] reported increased EROD activities (approximately 7-fold above reference activities) in ocean perch (*Perca fluviatilis*) at distances up to 8 km from Swedish bleached kraft mill effluent. Wild marine Dab (*Limanda limanda*) captured in the vicinity of oil platforms in the North Sea showed 5–10 fold EROD induction compared to fish caught 80 km away.[20] Van Veld et al.[45] found higher MFO induction in the intestine, as opposed to the liver, of spot (*Leiostomus xanthurus*) exposed to PAH-contaminated sediments. MFO induction is common in a variety of wild fish exposed to PAHs from crude oil[36] and in fish exposed to high PCB, organochlorine pesticides, and PAHs in sediments.[45–48]

An inherent difficulty in the use of wild fish is the variability in MFO levels among individuals in a common fish population. This variability makes comparison between populations difficult. An example of variability is shown in a population of white sucker in the vicinity of an urban river surveyed over a two-year period (Figure 5.1). As well as a considerable deviation in values within each sex/seasonal group, the population demonstrated a seasonal and sexual variability in EROD-specific activity. Peak EROD levels, particularly in male fish, occurred at spawning, then declined, reaching lowest levels in the nonspawning fall period.

EROD levels may increase in fish populations as they proceed up their spawning stream and theoretically may be subjected to higher contaminant levels. If the spawning habitat is outside contaminated area, then EROD activities in spawning fish can decline. Temperature may be another factor that can influence MFO activity, as fish metabolic rate rises and falls with water temperatures. As reported by Koivusaari,[11] hepatic P-450 levels in male and female rainbow trout increased with water temperature. The seasonal variation in MFO activity seen with white sucker has similarly been reported with wild roach (*Lueciscus rutilus* L.) in which constitutive MFO activities peaked during summer months.[14]

Sex Differences

In the wild, EROD activity in male white sucker is generally higher than in female fish (Figure 5.1). Sex differences were also reported for rainbow trout and

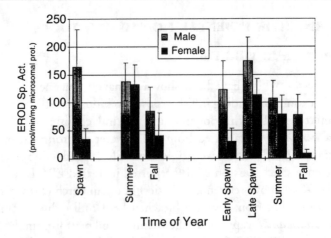

Figure 5.1. Seasonal and sexual deviation in EROD levels in wild common white sucker *Catostomus commersoni* captured over a two-year period in or near an urban spawning river.

brook trout, with P450 protein content in livers of males double those of female fish.[49] In white sucker, EROD activity in male and female fish may be similar during the summer. Sexual difference was most pronounced during spawning, when levels in female fish were markedly reduced from summer fish, perhaps due to hormonal regulation.

Although the measurement of MFO activity in wild fish is a useful tool for environmental monitoring, it is sometimes difficult to define critical parameters such as exposure history, diet, and general level of health. In addition, sex differences and seasonal effects (i.e., spawning) may contribute to the variation in enzyme activity. To better define changes in MFO activity resulting from exposure to industrial effluents or specific contaminants, MFO can be examined in transplanted caged fish. To further reduce variability seen in wild fish, a laboratory MFO bioassay offers many advantages.

Fish MFO Bioassays In Situ

In addition to reducing variation in MFO activity observed in resident fish populations, caged fish bioassays provide a relatively rapid means of assessing MFO induction potential in aquatic environments. This provides quick feedback and data for analysis or action. Caged fish exposure is relatively short (2–4 days) and measurements of EROD levels are rapid. The induction response appears to be relatively robust and not easily influenced by fish stress, or minor variations in light, temperature, or feeding. In contrast, bioassays measuring fish steroids and plasma cortisol or glucose levels can be dramatically influenced by season, natural biological rhythms, and handling stress.[50] It is feasible to perform caged fish bioassays for MFO induction in situ, at locations at the side of a river, in affected water bodies, or in an industrial site. As well, wild fish can be caged for short times in effluent outfalls or in gradient plumes.

Caged fish bioassays have been used to measure an increase in MFO activity in fish exposed to bleached kraft mill effluents. White sucker (*Catostomus commersoni*) caught at a reference site and caged for 3 days in bleached kraft mill effluent (BKME) showed 9- and 31-fold EROD induction compared to fish caged at the resident reference site (Figure 5.2). Fish were captured at the reference site during the spring spawning season. The caged fish bioassay permitted a study of MFO induction by BKME in fish in the stage of spawning. Constitutive MFO levels in male fish, at the reference site, were slightly higher (1.5 times) than constitutive levels in females. Increases in EROD levels in male fish exposed to BKME also exceeded that of exposed female fish. It was assumed that the control of MFO induction by sex hormones in spawning female fish may dampen induction in exposed fish. While spawning hormonal conditions may influence MFO activity, it clearly does not cause a total inability of female fish to respond to external MFO inducers.

Control EROD activities of caged wild fish were similar to those of wild fish populations at the reference site, showing little or no influence of the transport and caging on the levels of EROD activity. Caging of wild fish allows experimenters to control the exposure by positioning fish in the effluent plumes at chosen concentrations. "Naturally" exposed white sucker showed 4-fold (males) to less than 1.5-fold (females) induction in the spring,[51] presumably because the spawning stream is distant from the area of the bay impacted by the BKME effluent. Caging spawning fish in the effluent plume (a stream of 25–50% effluent) resulted in a dramatic induction, even in female fish which appear relatively recalcitrant to inducers.

Laboratory MFO Bioassays

Laboratory bioassays use a standardized reference fish species, usually the rainbow trout (*Oncorhynchus mykiss*) originating from a common fish stock and grown under controlled laboratory conditions where diet (lacking MFO inducers) and holding conditions are strictly controlled. Theoretically this would reduce the variation in constitutive MFO levels between fish. Sexually immature fish are used, thus reducing sexual differences in MFO levels. Fish are exposed under controlled laboratory conditions where only the test effluent or test chemi-

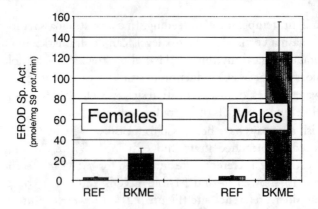

Fish caged for 3 days in REF or BKME streams

Figure 5.2. Geometric mean white sucker hepatic EROD activity in fish caged in BKME and at a clean site. Fish were captured at the clean site by trap netting in a spring 1996 spawning run, and held for 3 days in a BKME-discharge stream or in a reference stream.

cal are responsible for MFO induction. A dilution water control or carrier control is included in all tests. Control fish allow a measure of variance in constitutive MFO levels as a result of varying laboratory conditions. Control fish also provide the best reference for use in calculating MFO induction. Measurement and reporting of relative levels of induction between different effluents or between test chemicals is valid, because of the stability of laboratory fish as verified by the laboratory control.

EROD Induction by 2378-TCDD

Rainbow trout dosed with TCDD show very high EROD activities nearly 200 times those of control fish. EROD activity rose in a graded fashion in response to increasing oral doses of TCDD (Figure 5.3). EROD activities of 200–800 pmole/mg S9 protein/minute are common in response to TCDD and other 2378-substituted dioxins or furans,[4] but are very rare with exposure to environmental effluents. This high induction by dioxins and furans is typical of these compounds, which are potent inducers and difficult compounds to metabolize. Most environmental inducers appear to be easily metabolized, and induction is weaker than observed for dioxins and furans. Usually, exposure to PAHs, exposure to undiluted bleached kraft mill effluent, or exposure to oil refinery effluent causes 10–50 fold MFO induction over control levels.

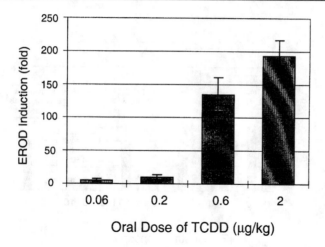

Figure 5.3. Mean TCDD-dosed rainbow trout hepatic EROD activity expressed as fold induction over control fish (EROD = 1.96 pmole/mg S9 prot./min). Fish were dosed orally with TCDD (0.060 to 2.0 mg/kg body weight) in gelatin capsules. EROD activities were measured after 2–16 days exposure. Error bars represent ± one standard error.

EROD Induction by BNF

Small rainbow trout exposed to 10 mg BNF/L at 15°C in the laboratory usually show about 10- to 20-fold induction over control fish (Figure 5.4). Constitutive EROD activities in rainbow trout are typically low, <1–2 pmol/mg S9 prot./min, and can rise to 50–100 pmole/mg S9 prot./min in response to PAH-type inducers, or with exposure to industrial effluents.

Other species of fish can be obtained from stock cultures and tested for EROD in the lab (Figure 5.4). The robustness of the EROD induction in fish permits bioassays on site under harsh winter conditions as long as fish are acclimated. Small chinook salmon (*Oncorhynchus tshawytscha*) and coho salmon (*Oncorhynchus kisutch* [Walbaum]) were induced by exposure to 1–100 μg/L β-naphthoflavone even under winter conditions (0–1°C). Chinook had higher constitutive levels of EROD, and were induced to higher absolute levels, but both species peaked at about 3-fold induction over controls (10 μg BNF/L). Higher concentrations of BNF appeared slightly toxic or inhibitory to the enzyme, as EROD activities at 100 μg BNF/L were significantly lower than those at the peak of 10 μg/L.

Fish captured in the wild can add another dimension to EROD studies. Tolerant species can be brought into the lab for short exposures, to study their response to known inducers, or to controlled concentrations of effluents. Peamouth chub (*Mylocheilus caurinus* [Richardson]) caught in the wild and exposed on site

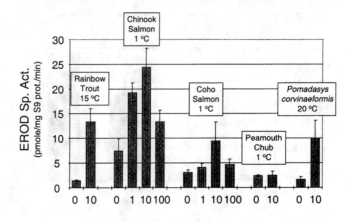

BNF Exposure Concentration (µg/L)

Figure 5.4. Geometric mean hepatic EROD activity of rainbow trout, chinook salmon, coho salmon, peamouth chub, and *Pomadasys corvinaeformis* exposed for 4 days to waterborne BNF. Exposures were performed at various temperatures, to which the fish were previously acclimated. Trout, chinook, and coho were cultured fish exposed in the lab or in situ. Peamouth chub and *Pomadasys* were captured wild fish exposed in the lab. Error bars show ± one standard error.

in winter (0–1°C) were not induced by 10 µg/L BNF. But a marine warmwater species, *Pomadasys corvinaeformis,* caught off the eastern coast of Brazil showed 6-fold EROD induction when exposed to BNF (10 µg/L) solutions. Such exposures to BNF and other inducing compounds can determine the range of MFO induction and whether the species is an appropriate biomonitor for effluent or contaminant exposure.

EROD Induction by Effluents in Laboratory Bioassays

Laboratory MFO bioassay can be applied to the testing of industrial effluents. Fish exposed to final effluents from pulp mills employing various types of processes have shown induction from 2- to 39-fold in exposures as short as 4 days.[52-54] Some mill effluents (maximum fish exposure concentrations tested = 10% effluent) showed no MFO induction, and induction varied with mill type.

MFO induction occurred in fish exposed to petroleum or crude oil.[17,19,20] We investigated the MFO-inducing ability of several Ontario refinery effluents. EROD induction was detected in all four refinery effluents at levels comparable to or higher than induction observed for BNF (10 µg/L).[55]

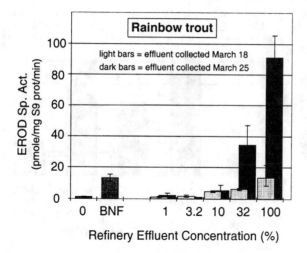

Figure 5.5. Geometric mean rainbow trout hepatic EROD activity after 4-day exposures to refinery effluent or β-naphthoflavone (BNF). Error bars show ± one standard error.

Refinery effluent was a potent inducer of EROD, with fish exposed to 100% refinery effluent showing 10- to 60-fold MFO induction (Figure 5.5). Fish exposed to refinery effluent collected on two dates, March 18 and March 25, showed large differences in EROD activity. This suggested the effluent was quite variable, and contained significantly more MFO inducers (probably PAHs) on March 25 compared to March 18. There was no difference in the storage or temperature of the collected effluent, and less than one day passed between time of effluent collection and the beginning of fish exposures. Control fish EROD and positive control (BNF-exposed) fish EROD remained consistent between bioassays, so the changes could not be ascribed to altered fish sensitivity.

Testing of Samples with Unknown EROD Induction Potential

A bioassay for EROD induction, in this case in conjunction with the rainbow trout 96-hour acute lethality test, may be used in studies of environmental samples with unknown induction potential. Applied to a study of highway runoff, the combined bioassays found samples to be occasionally acutely lethal to rainbow trout and generally capable of inducing hepatic EROD levels in fish at nonlethal sample dilutions.[56] In addition to highway runoff, stormwater originating from highway/commercial sources (at the inlet to a stormwater retention pond) as well as snow collected at a highway curb were capable of EROD induction (Figure 5.6).

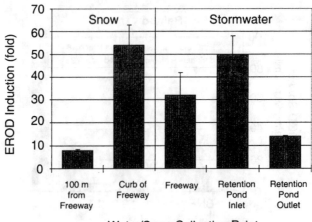

Figure 5.6. Mean rainbow trout hepatic EROD activity (pooled livers of ten fish exposed for 4 days) induced by multiple samples of roadway snow and stormwater. Results expressed in fold induction referenced to EROD levels (3.58 ± 1.17 pmoles/min/mg microsomal protein) in fish held in control dilution water. Error bars show ± one standard deviation, and represent the variability in induction by different samples.

The bioassay was also useful in defining the extent of EROD induction potential surrounding a contaminant source. For example, samples of snow at the highway curb were capable of EROD induction, while samples of snow collected 100 m from the roadway were much less potent. The bioassay was used to examine the effectiveness of stormwater treatments in reducing EROD induction potential. Levels of EROD induction in stormwater could be reduced (compared to EROD induction by inlet and outlet samples) by the temporary holding of stormwater in a retention pond (Figure 5.6).

A laboratory bioassay for EROD activity in rainbow trout may be used to better define possible sources of active contaminants. Since highway runoff induced EROD activity in rainbow trout, the source of contaminants may be vehicles or the fluids used in normal operation of these vehicles. A study of a limited number of vehicle fluids found EROD induction potential in aqueous extracts of diesel fuel, waste motor oil from both gasoline and diesel engines, and in rubberized under-coating (Figure 5.7). In contrast, radiator antifreeze, windshield washer fluid, brake fluid, power steering fluid or hydraulic fluid contained minimal or no EROD induction potential. Although aqueous extracts of motor oil contained minimal induction potential, an extract of waste oil from a gasoline engine was more potent than waste oil from a diesel engine.

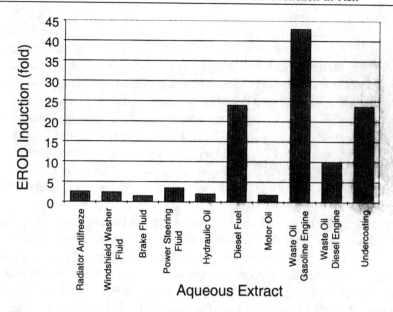

Figure 5.7. EROD induction in rainbow trout (10 pooled livers) exposed for 4 days to solutions or aqueous extracts of vehicle fluids. Results expressed in fold induction referenced to EROD levels (2.26 ± 0.88 pmoles/min/mg microsomal protein) in fish held in control dilution water.

The detection of EROD induction potential in waste motor oil was consistent with the results of Upsall et al.[57] Similarly, exposure to crude oil has been shown to induce MFO levels in marine codfish, sculpin, and flounder.[17] In the study of vehicle fluids, all possible sources of MFO-inducing substances associated with vehicular traffic were not explored. The demonstration of EROD induction by petroleum related vehicle fluids sets priority for further studies identifying EROD-inducting compounds in highway runoff.

Further application of bioassays for MFO induction in fish may be of use in the identification and characterization of compounds associated with EROD induction in environmental samples. In the case of the study of vehicle fluids, the specific compounds associated with EROD induction were not identified, although some of these compounds were extractable from petroleum based fluids into water. Polyaromatic compounds were a suspected source of MFO induction by waste crankcase oil.[57] However, more detailed studies are still required to identify the compound(s) associated with EROD induction by highway runoff or petroleum based vehicle fluids.

The rainbow trout EROD induction bioassay has been used in toxicity identification and evaluation (TIE) of effluents and chemical formulations. In one study, EROD induction in rainbow trout was used to identify compounds in thermomechanical pulp mill effluent.[54] EROD induction by the effluent was traced to the thermomechanical condensate (TMPC) wastewater. Chemical extraction and chromatographic techniques, along with a trout MFO bioassay, were use to identify juvabione and dehydrojuvabione as the active compounds. In a second study, EROD induction in rainbow trout, along with solid phase extraction and reverse-phase high pressure liquid chromatography (HPLC) techniques, were used to characterize EROD-inducing contaminants in a formulation of the lampricide 3-trifluoromethyl-4-nitrophenol (TFM).[58] The study found that dioxin-like contaminants formed during synthesis of TFM were the likely inducing compounds.

Although EROD induction in rainbow trout has been successfully applied in TIE evaluations, a practical limitation may the amount of test material or compound required by the bioassay. Trout are exposed in water and require a suitable loading density (calculated on the weight of fish X days of exposure/volume of water). Since considerable volumes of water are involved, several grams of purified material may be required. In the final stages of a TIE study, researchers may decide to apply more sensitive and miniature bioassays, such as MFO induction in cell or tissue cultures, to complete compound identification.

Advantages and Disadvantages of Various Fish Exposure Techniques

MFO induction in wild fish, in caged fish and in lab-exposed fish can be a useful bioindicator, and each of the assessment approaches has its own merits and drawbacks (Table 5.2). Historically, the assessment of MFO in fish has progressed from studies of wild fish, to encompass caged fish and lab-exposed fish.

MFO in wild fish is often helpful in assessing the exposure history of the fish, to determine if fish have been in a contaminated area for the past few days. MFO induction is rapid, so elevated EROD activities reflect the immediate past exposure of the wild fish. Problems associated with the measurement of MFO in wild fish stem mainly from the natural variability of the fish, and the biological rhythms associated with time of year, temperature, and the sex and spawning state of the fish.

The capture and caging of wild fish in a contaminated area can provide a good comparison point to assess MFO levels in wild uncaged fish and to determine the inducibility of an unfamiliar species. Caging of standard laboratory species is another way to assess MFO-inducing ability of effluents or contaminated areas. This approach allows MFO in caged fish to be compared site-to-site, and allows an approximate ranking of sites and effluents.

Table 5.2. Advantages and Disadvantages of Assessment of MFO Activities in Wild Fish, Caged Fish or in Fish Exposed in the Laboratory.

Advantages	Disadvantages
MFO in Wild Fish	
• Fish exposed to natural environment • Unnatural stresses limited to capture • Fish undergo natural development and reproductive cycles • Fish show integrated response to the accumulated natural and contaminant stresses	• Unknown exposure history, and potential avoidance behavior • Known and unknown variables (including natural hormonal, seasonal, temporal, nutritional, and genetic variability) may influence fish response • High cost of field surveys
MFO in Caged Fish	
• Fish exposed to a component of the natural environment • Fish movement is limited and exposure is better defined • Control over duration and level of exposure • Control over fish species, sex, age, and life-stage exposed • Studies using caged cultured fish may allow an approximation of site-to-site differences	• Fish are stressed by transfer, containment, and capture • Captured wild fish are influenced by natural seasonal, hormonal, and genetic variability • Cultured fish may not adequately represent native fish populations • Intermediate in cost. However, if wild fish are captured for caging, costs can be as high as field surveys
MFO in Fish Laboratory Bioassay	
• Fish exposed under controlled conditions with standardized test protocols • Small juvenile animals used; therefore, sexual, hormonal, and seasonal variability is minimized • Fish exposed to a defined range of contaminant concentrations • Testing of effluents/chemicals under standardized species/ conditions makes comparisons between sites possible • Standardized bioassay allows toxicity identification and evaluation (TIE) of complex effluents • Relatively low cost	• Controlled short exposures cannot mimic long-term duration and/or variability of exposure in the natural environment • Natural environmental/biological variables which normally affect fish response are not considered • Fish species used in test protocol may not be representative of species in the natural environment

Laboratory fish exposures provide the highest degree of control over the fish and the effluent. The use of small, immature species minimizes the potential influence of age, stage and reproductive cycle on MFO activity. Exposures can be run in the laboratory year-round, allowing the assessment of seasonal differences in effluents.

Regardless of the species studied and fish exposure method employed, MFO can provide an indication of a fish's exposure to potentially harmful lipophilic xenobiotics. Beyond that, the significance of MFO induction is a widely debated subject.

SIGNIFICANCE OF MFO

MFO is a metabolic process common to a wide variety of animal species. Its primary function is detoxification and excretion of lipophilic endogenous and exogenous compounds. MFO is under biological control, and levels of these enzymes naturally fluctuate to allow an organism to respond to a changing internal and external environment. The question arises whether elevation of MFO enzymes is a sign of deterioration of the health of the animal or whether it is a normal biological response. It is widely debated whether exposure to contaminants which result in an excessively high or prolonged elevation in activities of MFO enzymes is a detrimental effect.

A possible detrimental effect of increases in MFO is the potential to convert promutagens into their active mutagenic forms. Elevated MFO activity potentially hastens this process. Short-term genotoxicity bioassays, for example the Ames test,[59] use an artificial system containing elevated levels of phase I MFO enzymes, to metabolically activate promutagens such as benzo[a]pyrene. Inclusion of a MFO activation system facilitates detection of promutagenic compounds and increases the sensitivity of mutagenicity tests. In vitro studies show that activation of promutagen with MFO enzymes can dramatically increase the efficacy of these mutagens. Whether the same process occurs in vivo, in organisms with artificially high MFO, has not been proved. Evidence against this concern is the absence of a demonstrated difference in incidence of liver neoplasia between fish populations distinguished only by differences in induced MFO levels.

Field studies of white sucker populations exposed to BKME often show elevated MFO activity. These fish also had decreased circulating steroid levels and a series of physiological disturbances, such as enlarged liver, reduced gonad growth, and delayed sexual maturity.[13,39] Lehtinen[60] suggests two hypothetical consequences of increased MFO activity: A primary pathway involving the excess formation of free radicals (through elevated levels of oxidized metabolites), and a secondary pathway involving an imbalance of steroid hormones. However, no cause-effect relationship could be drawn between increased MFO activity and decreased levels of steroids in blood of fish exposed to BKME.[44]

Increased MFO activity is a detoxification response that signals an organism's exposure to potentially toxic compounds. While the purpose of MFO is detoxification and protection of the organism, persistent organic anthropogenic chemicals may result in a significantly elevated level of these enzymes over a prolonged period of time. At the very least, a permanently elevated MFO system may divert biological substrates and energy into detoxification processes at the expense of growth and reproduction.

While there is little data to directly support a mechanistic link between MFO and other health effects, there is also a lack of data to refute this link. Payne et al.[36] probably sum it up best, concluding that MFO induction is a primary detoxification response to potent inducing chemicals and "enzyme induction can often serve as an early warning signal of possibly more serious pathologies."

CONCLUSIONS

MFO induction is a sensitive tool to measure exposure of fish to potentially toxic contaminants from a variety of anthropogenic sources. EROD induction has been used in wild fish populations to measure exposure and to define areas affected by industrial effluents. Fish caging, in situ and laboratory exposures can more efficiently rank inductive potencies of effluents, and can identify sites for more detailed biological investigations. Sensitive laboratory bioassays can be designed to more efficiently detect induction in environmental samples, and to compare potencies among discharges. Laboratory bioassays are most useful in toxicity identification and evaluation, and for the detection of sources, identities, and potencies of inducing compounds.

REFERENCES

1. Martin, D.W., Jr., P.A. Mayes, V.W. Rodwell, and D.K. Granner. *Harper's Review of Biochemistry*, 20th ed., Lange Medical Publications, Los Altos, CA, 1985, p. 132.
2. Greenlee, W.F. and A. Poland. TCDD: A Molecular Probe for Investigating the Induction of Aryl Hydrocarbon Hydroxylase, in *Monographs of the Giovanni Lorenzini Foundation, Vol. 1. Dioxin: Toxicology and Chemical Aspects,* Cattabeni, F., A. Cavallaro, and G. Galli, Eds., SP Medical & Scientific Books, New York, 1978, p. 113–122.
3. Bradlaw, J.A. and J.L. Casterline. Induction of enzyme activity in cell culture: A rapid screen for detection of planar polychlorinated organic compounds. *J. Assoc. Off. Anal. Chem.,* 62(4), pp. 904–916, 1979.

4. Parrott, J.L., P.V. Hodson, M.R. Servos, S.L. Huestis, and D.G. Dixon. Relative potency of polychlorinated dibenzo-*p*-dioxins and dibenzofurans for inducing mixed function oxygenase activity in rainbow trout. *Environ. Toxicol. Chem.*, 14, pp. 1041–1050, 1995.

5. Okey, A.B. Enzyme induction in the cytochrome P-450 system. *Pharmacol. Ther.*, 45, pp. 241–298, 1990.

6. Poland, A. and J.C. Knutson. 2,3,7,8-Tetrachlorodibenzo-*p*-dioxin and related halogenated aromatic hydrocarbons: Examination of the mechanism of toxicity. *Ann. Rev. Pharmacol. Toxicol.*, 22, pp. 517–554, 1982.

7. Nebert, D.W. and D.R. Nelson. In *Methods of Enzymology, Cytochrome P-450*, Vol 206, Waterman, M.R. and E.F. Johnson, Eds., Academic Press, Orlando, FL, 1991, pp. 3–11.

8. Poland, A. and E. Glover. Chlorinated dibenzo-*p*-dioxins: Potent inducers of d-aminolevulinic acid synthetase and aryl hydrocarbon hydroxylase. II. A study of the structure-activity relationship. *Molec. Pharmacol.*, 9, pp. 736–747, 1973.

9. Gillette, J.R., D.C. Davis, and H.A. Sasame. Cytochrome P-450 and its role in drug metabolism. *Ann. Rev. Pharmacol.*, 12, pp. 57–84, 1972.

10. Lu, A.Y.H. Liver microsomal drug-metabolizing enzyme system: functional components and their properties. *Fed. Proc.*, 35, pp. 2460–2463, 1976.

11. Koivusaari, U. Thermal acclimatization of hepatic polysubstrate monooxygenase and UDP-Gluconosyltransferase of mature rainbow trout (*Salmo gairdneri*). *J. Exptl. Zool.* 227, pp. 35–42, 1983.

12. Smith, I.R., C. Portt, and D.A. Rokosh. Hepatic mixed function oxidases are induced in populations of white sucker, *Catostomus commersoni*, from areas of Lake Superior and the St. Mary's River. *J. Great Lakes Res.*, 17(3), pp. 382–393, 1991.

13. Munkittrick, K.R., C.B. Portt, G.J. Van Der Kraak, I.R. Smith, and D.A. Rokosh. Impact of bleached kraft mill effluent on population characteristics, liver MFO activity, and serum steroid levels of a Lake Superior white sucker (*Catostomus commersoni*) population. *Can. J. Fish. Aquat. Sci.*, 48, pp. 1371–1380, 1991.

14. Dewaide, J.H. and P.Th. Henderson. Seasonal variation of hepatic drug metabolism in the roach, *Leuciscus rutilus* L. *Comp. Biochem. Physiol.*, 32, pp. 489–497, 1970.

15. Pensrose, W.R., R.G. Murphy, L.L. Dawe, W.D. White, W.P. Gulliver, and D.G. Walton. Transformation of 2,5-diphenyloxazole and derivatives by mixed-function oxidases of a marine fish. *Chemosphere*, 8, pp. 509–520, 1979.

16. Binder, R.L. and J.J. Stegeman. Induction of aryl hydrocarbon hydroxylase activity in embryos of an estuarine fish. *Biochem. Pharmacol.*, 29, pp. 949–951, 1980.

17. Payne J.F and L.L. Fancey. Effect of long term exposure to petroleum on mixed function oxidase in fish: Further support for use of the enzyme system in biological monitoring. *Chemosphere*, 11(2), pp. 207–213, 1982.

18. Spies, R.B., J.S. Felton, and L. Dillard. Hepatic mixed-function oxidases in California flatfish are increased in contaminated environments and by oil and PCB ingestion. *Marine Biol.,* 70, pp. 117–127, 1982.

19. Walton, D.G., L.L. Fancey, J.M. Green, J.W. Kiceniuk, and W.R. Penrose. Seasonal changes in aryl hydrocarbon hydroxylase activity of a marine fish *Tautogolabrus adspersus* (Walbaum) with and without petroleum exposure. *Comp. Biochem. Physiol.,* 76C(2), pp. 247–253, 1983.

20. Stagg, R.M., A. McIntosh, and P. Mackie. Elevation of hepatic monooxygenase activity in the dab (*Limanda limanda* L.) in relation to environmental contamination with petroleum hydrocarbons in the northern North Sea. *Aquat. Toxicol.,* 33, pp. 245–264, 1995.

21. Zijlstra, J.A. and E.W. Vogel. Influence of inhibition of the metabolic activation on the mutagenicity of some nitrosamines, triazenes, hydrazines and seniciphylline in *Drosophila melanogaster. Mutation Res.,* 202, pp. 251–267, 1988.

22. Lee, R.F., S.C. Singer, and D.S. Page. Responses of cytochrome P-450 systems in marine crab and polychaetes to organic pollution. *Aquat.Toxicol.,* 1, pp. 355–365, 1981.

23. Ingram, A.J. The lethal and hepatocarcinogenic effects of dimethylnitrosamine injection in the newt *Triturus helveticus. Br. J. Cancer,* 26, pp. 206–215, 1972.

24. Khudoley, V.V. and J.J. Picard. Liver and kidney tumors induced by N-nitrosdimethylamine in *Xenopus borealis* (Parker). *Int. J. Cancer,* 25, pp. 679–683, 1980.

25. Khudoley, V.V. The induction of tumors in *Rana temporaria* with nitrosamines. *Neoplasm,* 24(3), pp. 249–251, 1977.

26. Harshbarger, J., G. Cantwell, and M. Stanton. Effects of N-Nitrosodimethylamine in the Crayfish *Procambarus clarki,* in *Proceedings of Fourth International Coll. on Insect Path. Soc. for Invert. Path.,* 1971, pp. 425–430.

27. Khudoley, V.V. and O.A. Syrenko. Tumor induction by N-nitroso compounds in bivalve mollusks *Unio pictorium. Cancer Letters,* 4, pp. 349–354, 1978.

28. Rogers, I.H., C.D. Levings, W.L. Lockhart, and R.J. Norstrom. Observations on overwintering juvenile chinook salmon (*Oncorhynchus tshawytscha*) exposed to bleached kraft mill effluent in the upper Fraser River, British Columbia. *Chemosphere,* 19, pp. 1853–1868, 1989.

29. Hodson, P.V., M. McWhirter, K. Ralph, B. Gray, D. Thivierge, J.H. Carey, G. Van Der Kraak, D.M. Whittle, and M.C. Levesque. Effects of bleached kraft mill effluent on fish in the St. Maurice River, Quebec. *Environ. Toxicol. Chem.,* 11, pp. 1635–1651, 1992.

30. Kloepper-Sams, P.J. and E. Benton. Exposure of fish to biologically treated bleached-kraft effluent. 2. Induction of hepatic cytochrome P4501A in Mountain whitefish (*Prosopium williamsoni*) and other species. *Environ. Toxicol. Chem.,* 13(9), pp. 1483–1486, 1994.

31. Jimenez, B.D., L.S. Burtis, G.H. Ezell, B.Z. Egan, N.E. Lee, J.J. Beauchamp, and J.F. McCarthy. The mixed function oxidase system of bluegill sunfish, *Lepomis macrochirus*: Correlation of activities in experimental and wild fish. *Environ. Toxicol. Chem.*, 7, pp. 623–634, 1988.

32. Kurelec, B., Z. Matjasevic, M. Rajavec, M. Alacevic, S. Britvic, W.E.G. Muller, and R.K. Zahn. Induction of benzo[a]pyrene monooxygenase in fish and the Salmonella test as a tool for detecting mutagenic/carcinogenic xenobiotics in the aquatic environment. *Bull. Environ. Contam. Toxicol.*, 21, pp. 799–807, 1979.

33. Pedersen, M.G., W.K. Hershberger, P.K. Zachariah, and M.R. Juchau. Hepatic biotransformation of environmental xenobiotics in six strains of rainbow trout (*Salmo gairdneri*). *J. Fish. Res. Board Can.*, 33, pp. 666–675, 1976.

34. Munkittrick, K.R., G.J. Van Der Kraak, M.E. McMaster, and C.B. Portt. Reproductive dysfunction and MFO activity in three species of fish exposed to bleached kraft mill effluent at Jackfish Bay, Lake Superior. *Water Poll. Res. J. Can.*, 27(3), pp. 439–446, 1992.

35. Haasch, M.L., R. Prince, P.J. Wejksnora, K.R. Cooper, and J.J. Lech. Caged and wild fish: Induction of hepatic cytochrome P-450 (CYP1A1) as an environmental biomonitor. *Environ. Toxicol. Chem.*, 12, pp. 885–895, 1993.

36. Payne, J.F., L.L. Fancey, A.D. Rahimtula, and E.L. Porter. Review and perspective on the use of mixed-function oxygenase enzymes in biological monitoring. *Comp. Biochem. Physiol.*, 86C, pp. 233–245, 1987.

37. James, M.O. and J.R. Bend. Polycyclic aromatic hydrocarbon induction of cytochrome P-450-dependent mixed-function oxidases in marine fish. *Toxicol. Appl. Pharmacol.*, 54, pp. 117–13, 1980.

38. Stegeman, J.J., R.M. Smolowitz, and M.E. Hahn. Immunohistochemical localization of environmentally induced cytochrome P450IA1 in multiple organs of the marine teleost *Stenotomus chrysops* (Scup). *Toxicol. Appl. Pharmacol.*, 110, pp. 486–504, 1991.

39. Larsson, A., T. Andersson, L. Förlin, and J. Hardig. Physiological disturbances in fish exposed to bleached kraft mill effluents. *Wat. Sci. Tech.*, 20(2), pp. 67–76, 1988.

40. Collier, T.K, S.V. Singh, Y.C. Awasthi, and U. Varanasi. Hepatic xenobiotic metabolizing enzymes in two species of benthic fish showing different prevalences of contaminant-associated liver neoplasms. *Toxicol. Appl. Pharmacol.*, 113, pp. 319–324, 1992.

41. Stien, X., C. Risso, M. Gnassia-Barelli, M. Romeo, and M. LaFaurie. Effects of copper chloride in vitro and in vivo on the hepatic EROD activity in the fish *Dicentrarchus labrax*. *Environ. Toxicol. Chem.*, 16(2), pp. 214–219, 1997.

42. Hodson, P.V., S. Efler, J.Y. Wilson, A. El-Shaarawi, M. Maj, and T.G. Williams. Measuring the potency of pulp mill effluents for induction of hepatic mixed function oxygenase activity in fish. *J. Toxicol. Environ. Health*, 49, pp. 101–128, 1996.

43. Bradford, M. A rapid and sensitive method for the quantitation of microgram quantities of protein utilizing the principal of protein-dye binding. *Anal. Biochem.*, 72, pp. 248–254, 1976.

44. Munkittrick, K.R., G.J. Van Der Kraak, M.E. McMaster, C.B. Portt, M.R. van den Heuvel, and M.R. Servos. Survey of receiving water environmental impacts associated with discharges from pulp mills. 2. Gonad size, liver size, hepatic EROD activity and plasma sex steroid levels in white sucker. *Environ. Toxicol. Chem.*, 13(7), pp. 1089–1101, 1994.

45. Van Veld, P.A., D.J. Westbrook, B.R. Woodin, R.C. Hale, C.L. Smith, R.J. Huggett, and J.J. Stegeman. Induced cytochrome P-450 in intestine and liver of spot (*Leiostomus xanthurus*) from a polycyclic aromatic hydrocarbon contaminated environment. *Aquat. Toxicol.*, 17, pp. 119–132, 1990.

46. Melancon, M., S.E. Yeo, and J.J. Lech. Induction of hepatic microsomal monooxygenase activity in fish by exposure to river water. *Environ. Toxicol. Chem.*, 6, pp. 127–135, 1987.

47. Monod, G., A. Devaux, and J.L. Riviere. Effects of chemical pollution on the activities of hepatic xenobiotic metabolizing enzymes in fish from the River Rhône. *Sci. Tot. Environ.*, 73, pp. 189–201, 1988.

48. Van der Oost, R., H. Heida, A. Opperhuizen, and N.P.E. Vermeulen. Interrelationships between bioaccumulation of organic trace pollutants (PCBs, organochlorine pesticides and PAHs) and MFO-induction in fish. *Comp. Biochem. Physiol.*, 100C, pp. 43–47, 1991.

49. Stegeman, J.J. and M. Chevion. Sex difference in cytochrome P-450 and mixed-function oxygenase activity in gonadally mature trout. *Biochem. Pharmacol.*, 29, pp. 553–558, 1980.

50. Jardine, J.J., G.J. Van Der Kraak, and K.R. Munkittrick. Capture and confinement stress in white sucker exposed to bleached kraft pulp mill effluent. *Ecotoxicol. Environ. Saf.*, 33, pp. 287–298, 1996.

51. Munkittrick, K.R. Environment Canada, Burlington, Ontario, personal communication, April 1998.

52. Gagné, Q.F. and C. Blaise. Hepatic metallothionein level and mixed function oxidase activity in fingerling rainbow trout (*Oncorhynchus mykiss*) after acute exposure to pulp and paper mill effluents. *Water Res.*, 11, pp. 1669–1682, 1993.

53. Martel, P.H., T.G. Kovacs, B.I. O'Connor, and R.H. Voss. A survey of pulp and paper mill effluents for their potential to induce mixed function oxidase enzyme activity in fish. *Water Res.*, 28, pp. 1835–1844, 1994.

54. Martel, P.H., T.G. Kovacs, B.I. O'Connor, and R.H. Voss. Source and identity of compounds in a thermomechanical pulp mill effluent inducing hepatic mixed-function oxygenase activity in fish. *Environ. Toxicol. Chem.*, 16(11), pp. 2375–2383, 1997.

55. Sherry, J.P., B. Scott, J.L. Parrott, P.V. Hodson, and S. Rao. The Sublethal Effects of Petroleum Refinery Effluents: Mixed Function Oxygenase (MFO)

Induction in Rainbow Trout, in *Proceedings of the 4th International Conference on Aquatic Ecosystem Health,* Coimbra, Portugal, Abstract C-41, May 14–18, 1995.

56. Rokosh, D.A., R. Chong-Kit, J. Lee, M. Mueller, J. Pender, and G.F. Westlake. Toxicity of Free-Way Stormwater, in Proceedings of the 23rd Annual Aquatic Toxicity Workshop, Calgary, AB. Fisheries and Oceans Canada. *Can. Tech. Rep. Fish. Aquat. Sci.,* No. 2144, pp. 151–159, 1997.

57. Upsall, C., J.F. Payne, and J. Hellou. Induction of MFO enzymes and production of bile metabolites in rainbow trout (*Oncorhynchus mykiss*) exposed to waste crankcase oil. *Environ. Toxicol. Chem.,* 12, pp. 2105–2112, 1993.

58. Hewitt, M.L., K.R. Munkittrick, I.M. Scott, J.H. Carey, K.R. Solomon, and M.R. Servos. Use of an MFO-directed toxicity identification evaluation to isolate and characterize bioactive impurities from a lampricide formulation. *Environ. Toxicol. Chem.,* 15(6), pp. 894–905, 1996.

59. Ames, B.N., J. McCann, and E. Yamasaki. Methods for detecting carcinogens and mutagens with the Salmonella/mammalian-microsome mutagenicity test. *Mutation Res.,* 31, pp. 347–364, 1975.

60. Lehtinen, K.-J. Mixed-function oxygenase enzyme responses and physiological disorders in fish exposed to kraft pulp-mill effluents: A hypothetical model. *Ambio.,* 19, pp. 259–265, 1990.

Chapter 6

Vitellogenin Induction in Fish as an Indicator of Exposure to Environmental Estrogens

J. Sherry, A. Gamble, P. Hodson, K. Solomon,
B. Hock, A. Marx, and P. Hansen

INTRODUCTION

International interest was stimulated by reports from the United Kingdom (U.K.) of female characteristics in fish taken from waters that receive effluents from sewage treatment plants or wool scouring mills.[1] The feminization effects were attributed to the possible presence of estrogenic chemicals in the ambient water. The term "environmental estrogen (EE)" is loosely used to describe chemicals in the environment that can cause estrogen-like effects. Such chemicals can act via several mechanisms including mimicry of the receptor binding properties of the natural hormone 17β-estradiol, alteration of estradiol:testosterone ratios by modulation of hormone synthesis or metabolism, or alteration of estrogen receptor levels.

Environmental estrogens are one part of a larger problem: the potential of environmental contaminants to disrupt the endocrine systems of diverse vertebrates. Evidence has emerged that development and reproduction of wildlife can be affected by exposure to chemicals that interfere, either directly or indirectly, with the endocrine system.[2] Many xenobiotic contaminants such as the pesticide DDT and the coplanar PCBs can act to disrupt the endocrine system. Abnormal levels or combinations of some naturally occurring chemicals, such as phytoestrogens, can also interfere with vertebrate endocrine systems. Human activities that cause high releases of such chemicals into the aquatic environment could be harmful. There has been speculation that hormonal disruption in many individuals of a species could lead to adverse effects at the population level. The scale and extent of many of the apparent population level effects that have been attributed to endocrine disruptors (EDCs) are controversial.[2–7] Much of that controversy shall eventually be resolved through investigative and hypothesis testing strategies that suggest, and later confirm or refute, cause-effect relationships between reproductive effects and suspect chemicals.

There are several well-documented cases to suggest links between impaired reproduction in vertebrates and exposure to EDCs. Accumulated residues of DDE, a degradation product of the insecticide DDT, have been associated with eggshell thinning in birds.[8] O,p′-DDT, an estrogen, has been linked to reproductive failure due to its estrogenic effects in bird populations.[9] An accidental spill of DDT in Lake Apopka (FL) was the suspected cause of impaired reproduction in alligators that coincided with a suite of endocrine related effects including disrupted hormone levels in males and females and decreased penis size in males.[10] Follow-up studies attributed the observed demasculinization effects to a buildup of the DDT degradation product p,p′-DDE in the alligators' body fat. p,p′-DDE is an antiandrogen and is thought to create an estrogenic environment in the alligators by skewing the effective ratio of endogenous estrogen:testosterone.[11] Additional studies by the same group, however, also showed that DDT and some of its metabolites such as o,p′-DDE and DDOH, which are known to occur in alligator body fat, can bind to the alligator estrogen receptor (ER), which suggests that the underlying mechanism for the alligators' reproductive disorders may be complex.[12]

Overindulgence in clover, rich in phytoestrogens, has been long known to cause reproductive impairment in ewes.[13] A recent review points out that phytoestrogens in the diet can inhibit the growth and formation of cancer in humans; receptor mediated processes and other mechanisms may be involved at the subcellular level.[14] Exposures to mixtures of halogenated planar aromatic hydrocarbons such as PCDDs, PCDFs, and coplanar PCBs have been linked to impaired reproduction in herring gulls and mink[15,16] and impaired development in fish.[17] Foster[18] concluded in a detailed review of the potential of xenobiotic contaminants of the Great Lakes to affect human reproductive processes, that the data were insufficient to prove the existence of a strong link to effects in the general population. A number of contaminants, however, that are part of the human body burden in the Great Lakes area, such as PCBs and DDT, may have the potential to affect sensitive individuals through additive or synergistic effects.[18] Clearly there is a need for better data and more research before firm conclusions or accurate risk assessments are possible.

A series of Canadian studies showed that white sucker (*Catostomus commersoni*) populations in an area of Lake Superior impacted by pulp mill effluent had smaller gonads, increased age to maturity, lower fecundity with age, altered secondary sex characteristics in males, and decreased levels of serum testosterone and estradiol.[19] Gagnon et al.[20] observed greater length at maturity, reduced gonad size, and more variable fecundity in populations of *C. commersoni* located downstream of a bleached kraft pulp mill on the St. Maurice River in Quebec. Kovacs et al.[21] recently reviewed Canadian investigations of the possible effects of pulp and paper mill effluent on the reproduction of wild fish, and recommended that future research pursue an integrative approach built on laboratory testing, mesocosm studies, and field investigations.

Bortone and Davis have associated masculinization effects in female mosquitofish (*Gambusia affinis*) and some other live-bearing species with exposure to bleached kraft mill effluent (BKME).[22] Phytosterols in the treated effluents were identified as the most likely cause of the effects. The authors concluded that females exposed to effluent for long periods would likely suffer impaired reproduction. The masculinization response, measured as anal fin elongation, has potential as a bioindicator of exposure to pulp mill effluents and has been proposed as a monitoring tool for assessing the benefits of improvements in mill effluent treatment on the recipient waters.[23] Many phytosterols, however, can act as estrogens, which raises the possibility that the ability to cause masculinization or feminization may depend on the balance of phytosterols in the final effluent or on other modifying factors.

We are interested in whether environmental estrogens are a problem in the Canadian aquatic environment. To address that issue we needed an indicator that could tell us, reliably and sensitively, whether organisms had been exposed to estrogenic substances. Researchers in England first proposed that induction of the egg yolk precursor vitellogenin (Vg) in male fish be used as a bioindicator of exposure to estrogenic chemicals.[24] Vg is normally produced in the liver of female teleosts following gene activation by estrogen receptors that have bound to the endogenous estrogen 17β-estradiol.[25] Vg is not normally found in high levels in male plasma, although low levels of Vg have been reported to occur.[26] The cause or function, if any, of low level Vg induction in males is not known at this time.

In the present chapter we focus on Vg as a bioindicator of exposure to EEs. After a brief introduction and overview of the supporting literature we describe our ELISAs for the measurement of Vg in brown trout and rainbow trout. We also provide details of the bioassays we use to assess the estrogenic potencies of target chemicals, complex mixtures, ambient waters, and industrial effluents. The bioassays are based on three exposure routes: i.p. injection, whole fish exposure using a static renewal protocol, and caged exposure to ambient water. We also summarize some experiments in which the bioassays were used to test pulp mill effluents, black liquor from the pulping process, and a known environmental estrogen 4-*tert*-octylphenol for the ability to induce Vg.[27]

BACKGROUND

Anglers in England were reputedly the first to report that male fish taken from sewage treatment lagoons had both male and female gonads. About 5% of roach (*Rutilus rutilus*) taken downstream of the discharge point of effluent from a sewage treatment plant were reported to be hermaphroditic.[28] It was logically suspected that a component of the sewage effluent had disrupted the sexual development of the fish. The term "feminization" has been coined to describe the development of female characteristics in genetically male organisms. Although the initial reports of "feminized fish" occurring in U.K. waters were greeted with some skepticism, they have stood the test of time.

Vitellogenin as a Bioindicator

A group of researchers centered at Brunel University hypothesized that the egg yolk precursor Vg, which in fish is normally found in appreciable levels only in females, could be used as a bioindicator of exposure to chemicals that cause feminization effects. Their proposal was consistent with the general bioindicator principle that the induction of abnormal and measurable characteristics that are mechanistically linked to the subcellular action of known chemicals or stressors can indicate exposure to exogenous chemicals or stressors sharing that mode of action. The bioindicator principle can be applied to many environmentally affected processes, including reproductive abnormalities caused by EDCs.

As a normal constituent of egg laying vertebrates, Vg has been the subject of a large body of work dealing with its regulation, characterization, and measurement. Vg is a large calcium-binding phosphoglycolipoprotein that is secreted by the liver into the blood for transport to the ovary where it is incorporated into the developing oocytes and forms the major egg yolk protein. Fish Vg varies widely in size over a range of about 350–600 kDa.[25] Vg is assumed to exist as multimeric protein, it readily forms aggregates, may complex with other molecules, and may also function as a carrier molecule for the transport of hormones or xenobiotic contaminants, many of which are lipophilic.[26] Vg may also be subject to proteolytic action while circulating in the blood, which could contribute to temporal variability in its structure and size.[26] Vg's ability to bind calcium in large amounts, which probably contributes to the development of the embryonic skeleton, is facilitated by the protein's high level of phosphorylation. Goldfish Vg has a protein phosphorous content of 0.79%[29] compared to brown trout's 0.58%.[30] Vg is produced in large quantities by the ripening female fish; levels as high as 100 mg.mL^{-1} and 10 mg.mL^{-1} have been reported for vitellogenic rainbow trout[31] and sole,[32] respectively. During vitellogenesis Vg is the major plasma protein, and can be measured in the mucus as a convenient means of sex determination.[33,34]

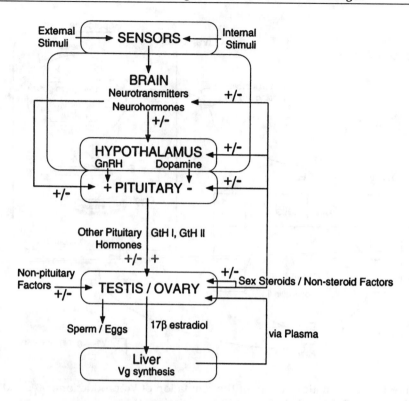

Figure 6.1. Schematic diagram of the hypothalamic-pituitary-gonadal-liver (HPGL) axis in fish. (Taken from Reference No. 96, with permission).

Vg production is under endocrine control. Figure 6.1 outlines the hypothalamic-pituitary-gonadal (HPG) axis which together with the liver regulates the production of Vg. The female reproductive hormone 17β-estradiol is synthesized and released from the ovarian tissue of female fish. The release of 17β-estradiol is stimulated by gonadotropin releasing hormone (GTH). GTH levels in turn are responsive to environmental signals and feedback regulation by various components of the HPGL which interact via positive and negative feedback loops. At the liver, estradiol activates the hepatocyte cells via a receptor-mediated process to trigger the production of Vg (Figure 6.2).[25] The receptor hormone complex forms a dimer and then binds to response elements to activate the downstream Vg genes.[35,36] The number of estrogen receptors in the liver also increases upon hormonal stimulation, which further increases the responsiveness and capacity of the system.[35] Once synthesized, Vg is transported in the blood stream to the ovaries where it is sequestered and cleaved to form the egg yolk proteins phosvitin and lipovitellin.

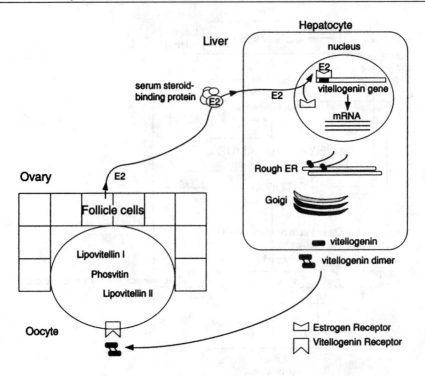

Figure 6.2. Schematic diagram of the regulation of Vg production by estradiol. (Taken from Reference No. 25, with permission).

Disruption of key processes, or loss of internal balance can follow environmental, mechanical, or chemical interference at multiple sites on the HPGL-axis. For example, Thomas demonstrated that Cd and Aroclor 1254 affected the HPGL axis in Atlantic croaker (*Micropogonias undulatus*) by apparently different mechanisms.[37] Cd caused ovarian growth to accelerate, and 17β-estradiol levels in plasma became elevated. In vitro secretion of gonadotropin (GTH) by pituitaries of Cd exposed fish was increased. On the other hand, Aroclor 1254 impaired ovarian growth and caused a decline in plasma estradiol. There was a corresponding decrease in the in vitro secretion of GTH by pituitary glands.[37] The vitellogenic response in brook trout (*Salvelinus fontinalis*) is known to be affected by acid stress.[38] Calcium-enriched freshwater has been shown to increase vitellogenin levels in rainbow trout.[39] The Vg control system is also sensitive to external and internal stimuli such as photo-periodicity and temperature changes.[40,96] Chen et al.[41] were among the first to consider a potential interaction between xenobiotic contaminants and the Vg cycle in fish. They reported that Aroclor 1254 and mirex depressed Vg, which they measured by rocket immunoelectrophoresis, in rainbow trout. Depression of Vg in female trout was proposed as an indicator of antiestrogenic effects.

The Brunel group tested Vg's value as a bioindicator by exposing trout to effluent at several sewage treatment plants. Vg in the plasma of exposed fish was measured by radioimmunoassay (RIA). Their initial hypothesis was confirmed when exposure of male or immature rainbow trout at several sites caused heavy induction (mg.mL^{-1}) of Vg.[24] Later studies showed that caged exposure of male or immature fish to sewage treatment plant effluents, or to sewage-impacted receiving waters from a variety of locations in the United Kingdom also caused the induction of Vg.[42,43] High levels of Vg induction were also reported for fish that were caged in the River Aire, whose waters are contaminated with alkylphenolic wastes from a wool scouring mill.[43] Several alkylphenols, which are degradation products of alkylphenol ethoxylates, can bind with low affinity to the estrogen receptor and trigger estrogenic responses.[44,45] Reproductive abnormalities, including the induction of Vg in males, and testicular abnormalities have been observed in flounder (*Platichthys flesus*) in English estuaries that are impacted by discharges from sewage treatment plants.[46]

While having diverse molecular structures, most estrogens contain a phenol, usually in the *p*-position, or a substituent group that can mimic the -OH group.[47] Most estrogens tend to be lipophilic and can bioconcentrate. Some environmental estrogens, such as methoxychlor, are bioactivated through the introduction of an -OH group. An important advantage of in vivo assays for the detection of estrogenicity is their general ability to bioactivate such "proestrogens." We shall briefly consider some classes of environmental chemicals, both naturally occurring and xenobiotic, that are estrogenic in fish.

p-Nonylphenol, a widely used antioxidant in the plastics industry, and a degradation product of nonylphenol polyethoxyate, a common nonionic surfactant, is known to be estrogenic to cultured mammalian cells, to rats and mice, and trout liver hepatocytes.[48–50] Moreover, NP binds competitively with mammalian ER.[49] Yeast cells that were transfected with the human ER and also contained an ERE reporter system have been used to demonstrate the estrogenicity of 4-NP and a range of other alkylphenol polyethoxylates and their degradation products.[51] The parent surfactants were nonestrogenic in the assay; however, many of the degradation products were weakly estrogenic. Both nonyl- (LOEC: 20.3 µg.L^{-1}) and octylphenol (LOEC: 4.8 µg.L^{-1}) have been shown to inhibit testicular growth and induce large-scale Vg production in male rainbow trout in a flow-through exposure.[52] Lech et al.[44] also exposed juvenile rainbow trout to nonylphenol (10–250 ng.L^{-1}) in a flow-through experiment. The estrogenic effects of the NP treatment were measured as Vg-mRNA which was detected by RT-PCR. For rainbow trout, NP was a far more potent estrogen (EC50: 14.14 ppb) than an acute toxicant (EC50: 194 ppb). NP can also induce the formation of testis-ova and altered sex ratios in Japanese medaka (*Oryzias latipes*) (LOAEC: nominal 50 μg.L^{-1} [ppb]).[53]

Nonylphenol also occurs in many sewage treatment plant effluents and in numerous rivers and estuaries in England and Wales[54] and in Canada.[97] Nonylphenoxyacetic acid (NP1EC) and related degradation products (NP2-4EC) of nonylphenol polyethoxlate surfactants have been detected in paper mill effluent in the United States.[55] It is not known whether pulp mill effluents that contain alkylphenols or their derivatives are estrogenic to fish. Several factors could influence the ability of alkylphenols in a complex mixture such as pulp mill effluent to induce Vg in fish. For example, the alkylphenol might not be biologically available, the response could be antagonized or blocked by other effluent components, or the alkylphenol could be biodegraded by microbes.

Exposure to waterborne doses of the broad-spectrum insecticide endosulfan was reported to depress plasma Vg in vitellogenic catfish (*Clarias batrachus*).[56] Donohoe and Curtis studied the possible estrogenic effects of the pesticides chlordecone, o,p'-DDT, and its degradation product o,p'-DDE in rainbow trout.[57] The pesticides were administered to the fish via i.p. injection, as was the 17β-estradiol control. Plasma Vg, which increased by 2400%, seemed to be a more sensitive indicator of estrogenicity than ER levels, which increased by 300% in liver hepatocytes. Chlordecone was reported to induce low levels of Vg in the trout. Both o,p'-DDT and o,p'-DDE were estrogenic in the trout; however, p,p'-DDE failed to induce Vg. The affinities of the pesticides for ER-binding sites in trout hepatic cytosol paralleled the Vg data.

QSAR considerations suggested to McKinney and Waller[58] that certain common hydroxylated metabolites of PCBs may have the ability to bind to the ER. Specifically, it was hypothesized that hydroxylated PCBs that are *o*-substituted, thus restricting their conformational freedom, could bind to the ER, which is a cleft-type receptor. That hypothesis was confirmed in in vitro assays using the MCF-7 cell line[59] and in receptor binding assays.[60] Pelissero et al.[61] showed that the following pure phytoestrogens could induce Vg in sturgeon: daizein, genistein, equol, coumestrol, and biochanin A. All were less potent than 17β-estradiol.

Effects of Induced Vg in Fish

There is little information available on the physiological, let alone the ecological consequences of Vg induction in male fish. The effects of exogenously induced Vg on both vitellogenic and nonvitellogenic females are also largely unknown. Herman and Kincaid[62] observed the following effects of exogenous 17β-estradiol (30 mg estradiol per kg feed for 76-days) on rainbow trout: enlargement of the liver, spleen, and kidneys, distention of the gas bladder, and dilation of the intestine with fluid. Massive deposits of an eosinophilic material, assumed to be Vg, were observed in the kidney, liver, and spleen. Diffuse deposits were observed in other areas of the body. Fifty percent of the treated fish died. The Vg deposits correlated with the observed fish mortality and may have been a contributory factor. There is a need for research into the possible health effects of exogenously induced Vg in individual fish as well as the consequences, if any, for the parent population.

ELISAs for Vg

In the past, the Vg content of fish plasma was deduced from estimates of alkali labile phosphate.[26,63] As a high molecular weight protein, Vg readily lends itself to analysis by immunological techniques. Antibody-based techniques have now become the methods of choice because of their excellent detection limits (DLs), responsiveness, low cost, ease of use, precision, accuracy, and rapidity. Radioimmunoassays (RIAs), which use radiolabeled Vg as the competing antigen, have been largely superseded by enzyme-linked immunosorbent assays (ELISAs), which eliminates the need to label purified Vg with a radiotracer. Apart from possible health and safety considerations, that is an important advantage since Vg is susceptible to damage during the radiolabeling procedure. The properties and performance characteristics of a selection of immunoassays for fish Vg are summarized in Table 6.1.

Other Assays for Environmental Estrogens

Other biological or biochemical tests can be used to assess the estrogenicity of pure chemicals and complex mixtures. Thomas and Smith used a cytosol suspension, which contained estrogen receptor, from the spotted sea trout (*Cynoscion nebulosus*) as the ligand binder in a competitive binding assay for the detection of estrogenic chemicals.[64] Kepone bound to the sea trout ER with lower affinity than to mammalian ER. Methoxychlor, o,p'-DDT, p,p'-DDT, o,p'-DDE, p,p'-DDE, and PCB mixtures failed to displace ³H-estradiol from the receptor binding sites, even though several of those compounds are estrogenic in mammals. O,p'-DDT and Aroclor 1254 (one of the PCB mixtures tested) can displace ³H-estradiol from mammalian ERs. Based on those observations, it is possible that some binding properties of ERs may be species specific. The authors concluded that mammalian assays or the use of mammalian ERs in either receptor binding assays or in in vitro assays may not accurately predict responses in fish.

The development of recombinant cell cultures has yielded several promising bioassays for the screening of chemicals and mixtures for estrogenic activity.[81]

Table 6.1. Descriptions and Performance Characteristics of Immunoassays for Fish Vg.

Species	Abs[a]	Cross Reactivity	Assay Type	Format	Working Range	DL[b]	Ref.[c]
Oncorhynchus kisutch (Coho salmon)	PABs	*O. spp.* and *S. clarki*	RIA	tube	50–1000 ng.mL^{-1}	50 ng.mL^{-1}	65
O. mykiss (rainbow trout)	PABs		RIA	tube	1–100 ng.mL^{-1}	1 ng.mL^{-1}	66
O. mykiss	PABs		ELISA	96-well plate; preincubation; equilibrium competitive binding; (Vg-Ab-AbE)		25 ng.mL^{-1}	57
O. mykiss	MABs	striped bass; other fish, amphibians, reptiles, birds	ELISA		1–80 µg.mL^{-1}	1 µg.mL^{-1}	67
O. mykiss	PABs		RIA	tube	5–100 ng.mL^{-1}	0.1 ng.tube^{-1}	31
O. mykiss	PABs	low cross-reactivity to Vg from other salmonids	ELISA	96-well: equilibrium competitive binding: (Vg-Ab-AbE)	33–1473 ng.mL^{-1}	9.4 ng.mL^{-1}	68
Salvelinus leucomaenis (whitespotted charr)	PABs		ELISA	96-well plate; sandwich (Ab-Vg-AbE)	10–100 ng.mL^{-1}	10 ng.mL^{-1}	69
Acipenser baeri (Siberian sturgeon)	PABs		ELISA	96-well plate; preincubation step; nonequilibrium competitive binding (Vg-Ab1-AbE)	10–1000 ng.mL^{-1}	15 ng.mL^{-1}	70

Species	Antibody	Notes	Assay	Assay details	Detection range	Detection limit	Ref
Morone saxatilis (striped bass)	PABs		radial-immuno-diffusion assay		4.5–100 µg.mL^{-1}	4.5 µg.mL^{-1}	71
M. saxatilis (striped bass)	PABs		ELISA	96-well plate; preincubation step; equilibrium competitive binding: (Vg-Ab1-AbE)	8–1000 ng.mL^{-1}	8 ng.mL^{-1}	33
Salmo salar (Atlantic salmon)	PABs		RIA	tube	0.25–30 ng	0.25 ng	72
S. trutta (brown trout)	PABs	rainbow trout and arctic charr (low sensitivity)	RIA	tube	10–100 ng.mL^{-1}	5 ng.mL^{-1}	73
Cyprinus carpio (carp)	PABs	most cyprinids	RIA	tube	2–200 ng.mL^{-1}	2 ng.mL^{-1}	74
Ictalurus punctatus (channel catfish)	MABs		ELISA	96-well plate; equilibrium competitive binding; (Vg-Ab-AbE)		15 µg.mL^{-1}	75
Cynoscion nebulosus (spotted sea trout)	PABs		RIA	tube	10–800 ng	10 ng	76
Dicentrarchus labrax L. (sea bass)	PABs		ELISA	96-well; preincubation; nonequilibrium competitive binding; (Vg-Ab-AbE)	1–60 ng.mL^{-1}	1 ng.mL^{-1}	77
Solea vulgaris (sole)	PABs		ELISA	96-well; preincubation; nonequilibrium competitive binding; (Vg-Ab-AbE)	10–800 ng.mL^{-1}	30 ng.mL^{-1}	32

Table 6.1. Descriptions and Performance Characteristics of Immunoassays for Fish Vg (continued).

Species	Abs[a]	Cross Reactivity	Assay Type	Format	Working Range	DL[b]	Ref.[c]
Anguilla japonica (Japanese eel)	PABs		ELISA	96-well; noncompetitive; (Ab-Vg-AbE)	10–10,000 ng.L^{-1}	0.8 ng.mL^{-1}	78
C. carpio (carp)	PABs	other cyprinids	RIA	tube	4–80 ng.mL^{-1}	4 ng.mL^{-1}	79
Sardinops melanostictus (Japanese sardine)	PABs		ELISA	96-well; noncompetitive sandwich assay; (Ab-Vg-AbE)	44–2760 ng.mL^{-1}	44 ng.mL^{-1}	80

a Antibodies.
b Detection limit.
c References.

The transfected cell lines contain a reporter gene which is located downstream of an estrogen response element (ERE). When the host cell's estrogen receptors are activated by a ligand, the reporter system triggers the reporter gene which usually encodes for a marker enzyme. The marker enzyme in turn generates a measurable signal. Such a recombinant cell line (MCF-7) was used to detect estrogenic activity in a sample of black liquor from the wood pulping process.[81] Le Drean et al.[82] used a transfected salmonid (STE-137) cell line to show that estrogen receptor (ER) from rainbow trout required 10 times more estradiol than ER from humans to achieve 50% of maximum activation. The specificity patterns of both trout and human receptors were the same. Those experiments provided additional evidence that test systems based on the human ER may not accurately predict effects in fish. Yeast cells containing an ERE-activated enzymatic reporter construct system have been transfected with cDNA for the human ER, and used to screen surfactants and their degradation products for estrogenicity.[51]

The E-screen test, based on the proliferative response of a human breast cancer cell line (MCF-7) to estrogens has been widely used to test environmentally important chemicals for estrogenic activity.[83] Other successful assays for the detection of estrogenicity have also been based on the proliferative response of MCF-7 cells.[59,98] Primary cultures of fish liver hepatocytes can also be used to screen chemicals and fractionated mixtures for the ability to induce Vg[50,84] and can provide valuable information when run in conjunction with whole fish tests.

BIOASSAYS FOR ENVIRONMENTAL ESTROGENS

Because our overall goal is to assess whether environmental estrogens are common in the Canadian aquatic environment, and whether they occur at levels that are likely to affect fish, our first objective was to develop tools to facilitate that assessment. The bioassays we use to screen samples and ambient waters for estrogenicity are based on the induction of Vg in brown trout and in rainbow trout. The induced Vg is measured by ELISA.

ELISAs for Trout Vg

We used brown trout from a natural stock that is maintained by the Ontario Ministry of Environment and Energy. Rainbow trout (60–100 g) were from Southern Ontario fish farms and were certified to be specific pathogen–free. Reagent Vg was induced in two year old male fish (100–400 g). The brown trout were held at 8–10°C, and the rainbow trout at 12.5–15°C with a 12-hour photoperiod for a minimum of two weeks prior to experiments.

Reagent Vg

Reagent Vg was raised in ten adult fish that were given a series of three weekly intraperitoneal (i.p.) injections (10 mg.kg^{-1}) of 17β-estradiol (Sigma, ON) dissolved in ethanol and diluted with 0.9% NaCl (w/v), after which their blood was harvested. The protease inhibitor aprotinin was added to the blood samples to minimize degradation of the Vg.[30] The blood plasma was separated by centrifugation (600 g for 15 minutes) and quick-frozen in liquid nitrogen. Vg was isolated from trout plasma by a triple precipitation procedure.[30,85] Four mL of 0.2M Na$_2$EDTA containing 4 TIU.mL^{-1} aprotinin was added to 1 mL of thawed plasma. After gentle mixing, 0.3 mL of 0.53 M MgCl$_2$ containing aprotinin (4 TIU.mL^{-1}) was added and the contents of the tube were again gently mixed. Ten mL of sterile distilled water was then added with further mixing, and Vg was allowed to precipitate for 15 minutes. The precipitate was harvested by centrifugation (1700 g for 15 minutes) and the resultant Vg pellet was dissolved in 1M NaCl containing aprotinin (4 TIU.mL^{-1}). The precipitation and harvesting procedure was repeated twice. The Vg pellet was then redissolved in 1 mL of the NaCl-aprotinin reagent, and dialyzed in two stages (Molecular Weight cutoff of 1,000).

The concentration of Vg in the stock solution was estimated from its absorbance at 280 nM.[73] The Vg stock was subsequently diluted with 50% glycerol and 0.05 M sodium phosphate buffer, pH 7.5, and stored in cryovials at –20°C. The addition of glycerol prevents freezing at –20°C, which can damage Vg.[73]

ELISA Outline

The ELISA for trout Vg, which is schematically represented in Figures 6.3 and 6.4, is a nonequilibrium competitive binding assay in which anti-Vg and analyte are preincubated together at 4°C for about 16 hours. That format is often used to enhance the sensitivity of immunoassays since it allows the antibodies to preferentially bind to the analyte molecules before they are exposed to the coating antigen. Purified Vg (coating antigen) is surface immobilized on the wells of a microtitration plate. Vg in solution competes with the bound Vg for antibody binding sites which are limited in number. The number of primary antibodies bound to the immobilized Vg is inversely proportional to the amount of Vg in solution. An enzyme-labeled second antibody is used to reveal the amount of bound primary antibody. Optical density is measured on a microplate reader, and the amount of Vg in samples is interpolated from a calibration curve.

ELISA Reagents

We describe the development and characterization of the ELISA for brown trout Vg in some detail and briefly summarize the performance characteristics of

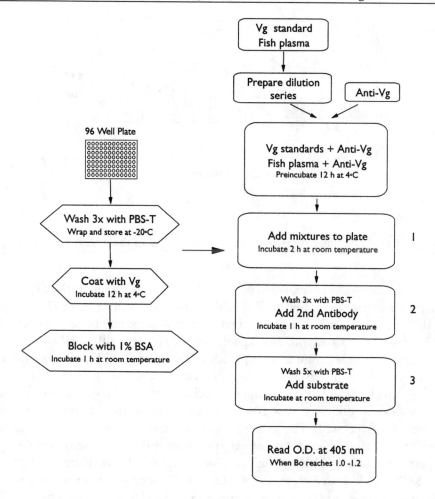

Figure 6.3. Flow chart outlining the ELISA for Vg. Numbers correspond to panels in the schematic diagram (Figure 6.4).

the ELISA for rainbow trout Vg. The key reagent in the brown trout ELISA is a polyclonal antiserum (anti-BtVg) that was raised in rabbits against brown trout Vg. The anti-BtVg was generously provided by Dr. B. Norberg of the University of Goteborg.[73] The antiserum was diluted 1 in 1000 in antibody diluent (100 mg bovine gamma globulin and 20 mg rabbit gamma globulin dissolved in 100 mL phosphate buffered saline (0.01M phosphate, pH 7.4, containing 850 mg NaCl)) and shared at –85°C. For the rainbow trout ELISA we used monoclonal antibodies (MABs) to rainbow trout Vg (anti-RtVg) that were generated in a collaboration among the authors under the auspices of the Canada-Germany Bilateral Agreement on Science and Technology.

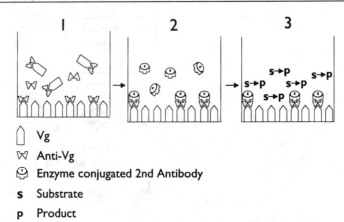

$\bigcap$ Vg
$\mathbb{W}$ Anti-Vg
$\mathfrak{Q}$ Enzyme conjugated 2nd Antibody
s Substrate
p Product

Figure 6.4. Schematic representation of key steps in the ELISA for Vg.

A checkerboard titration was used to select the working proportions of coating Vg and primary antibody. Twofold dilutions of purified BtVg (1,000 ng. mL⁻¹ to 31.25 ng.mL⁻¹) were prepared in 0.05M carbonate buffer, pH 9.6. A row of wells was coated with each Vg solution (200 μL per well), and then blocked with excess bovine serum albumin (BSA). The coating and blocking solutions were removed by inversion and the plate was washed three times with PBS-T (phosphate buffered saline containing 500 μL.L⁻¹ Tween-20). A dilution series of the anti-Vg in PBS-T (1:10 × 10³–1:320 × 10³) was added column-wise to the plate (200 μL per well). The plate was sealed and incubated at room temperature for 2 hours, after which it was washed three times with PBS-T. Second antibody (goat anti-rabbit IgG labeled with alkaline phosphatase), was diluted 1:2000 in PBS-T and added to each well (200 μL per well). After one hour at room temperature, the second antibody was removed and the plate was washed five times. The enzyme substrate (20 mg p-nitrophenyl phosphate) in 22 mL diethanolamine buffer (10% [v/v] diethanolamine in distilled water, pH 9.6) was then added to the wells (200 μL per well). The plate was gently agitated and then incubated at room temperature. The optical density of the wells was read at 30-, 45-, and 60-minutes at 405 nm in a multiwell plate reader. Combinations of coating-Vg and diluted anti-Vg that yielded an O.D. of 0.8 to 1.2 within 60 minutes were considered likely to yield workable ELISAs. Trial calibration curves were used to evaluate the performances of selected combinations. The trial curves had similar working ranges (20–500 or 1000 ng.mL⁻¹) and performance characteristics based on the 50- and 80% binding points. An antiserum dilution of 1:240,000 and 125 ng.mL⁻¹ coating-Vg was selected for use in the routine ELISA (Figure 6.5). The selection criteria were adequacy of the ELISA's working range, acceptable performance at low Vg concentrations, and conservation of key reagents.

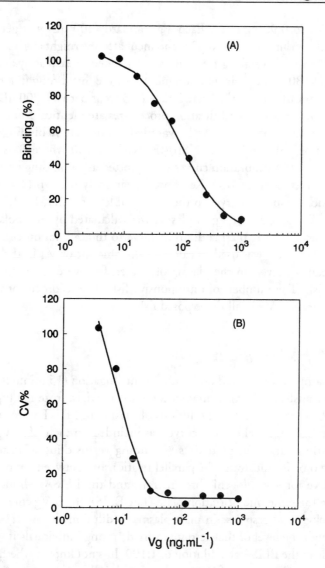

Figure 6.5. Calibration curve (a) and intra-assay precision profile (b) for the ELISA for brown trout Vg. The precision profile is derived from curve (a).

ELISA Procedure

The ELISA for Vg is outlined in Figures 6.3 and 6.4. The wells of a 96-well microtitration plate were coated with BtVg (125 ng.mL^{-1}); four uncoated wells were used to correct for nonspecific binding (NSB). Coated plates can be stored at 4°C until used. A 1,000 ng.mL^{-1} standard of Vg in PBS-T was diluted in a twofold series to 0.98 ng.mL^{-1}. Each Vg standard (400 µL) was gently mixed (1:1) with diluted anti-Vg (1:240,000) and incubated overnight at 4°C in a capped vial (12 × 75 mm). Several Vg-free vials were included in each assay as zero binding references (B0). The preincubation mixtures were then brought to room temperature and transferred to the coated plate (× 3 wells per vial; 200 µL per well). The ELISA and checkerboard titration procedures are identical from this point.

The mean O.D. of replicate wells was used to calculate % binding relative to the analyte free wells (B$_o$) as (B$_i$-NSB)/(B$_o$-NSB) × 100 where B$_i$ is the mean sample O.D. ELISA calibration curves were plotted as % binding vs. log$_{10}$ [Vg]. A four parameter Logistic Dose Response Equation ($y = a + b/(1 + (x/c)^d)$) was used to fit dose-response curves to the data (TableCurve 2D, Jandel Scientific, California). The amount of Vg in plasma was calculated by interpolating from the linear portion of the plasma dilution curve to the calibration curve. A dilution factor correction was used to calculate the amount of Vg in the undiluted plasma. Mean Vg levels in the plasma of treated fish were estimated from the responsive fish. The numbers of unresponsive fish were recorded for treatments that did not induce Vg in all the exposed fish.

Plasma Analysis

Blood samples were stored on ice until centrifugation (15 minutes at 600 g). Plasma was transferred from vacutainer tubes to cryovials, frozen in liquid nitrogen, and stored at –80°C. Serial dilutions of plasma in PBS-T were analyzed in triplicate by ELISA. A calibration curve was included on each ELISA plate. The plasma dilution curve was plotted as % binding versus dilution factor (Figure 6.6). Figure 6.6 also illustrates the parallel relationship between a typical plasma dilution curve for a vitellogenic brown trout and the ELISA calibration curve prepared on the same microtitration plate. The ELISA for BtVg consistently detected a background response in male plasma at dilutions below 1:100 (Figure 6.6). Although the level of that response varied among individuals it was always undetectable in the ELISA at a dilution of 1:100. It is not known whether the low level response in the plasma of untreated male fish is due to the presence of small amounts of Vg or to a cross-reacting plasma protein. Since male and female fish are held in the same tanks it is possible that the secretion of estradiol by female fish could induce low levels of Vg.

ELISA Precision

The intraassay variability of the ELISA, which was estimated from a heavily replicated (9 x) calibration curve, is presented in Figure 6.5. The CV was < 10%

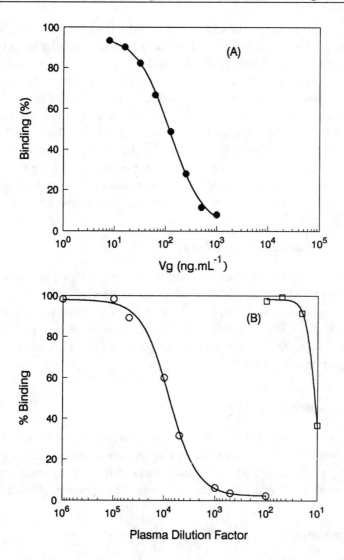

Figure 6.6. ELISA calibration curve (a) and fish plasma dilution curve (b) for brown trout Vg. Both curves were prepared on the same plate. □, diluted male plasma; ○ diluted plasma from estradiol induced female.

from 30–1000 ng.mL⁻¹ Vg. The ELISA's sensitivity, defined as the imprecision of measurement of zero dose,[86] was estimated to be 10.5 ng.mL⁻¹. The ELISA's practical detection limit (PDL), defined as two times the imprecision of the zero dose measurement, was estimated to be 21 ng.mL⁻¹. The PDL estimates the minimum amount of Vg that can be routinely measured in the ELISA more conservatively than the statistical estimate of ELISA sensitivity. Because the ELISA detects a response in plasma from untreated males, fish plasma samples must be diluted below this threshold (1:100) to avoid possible interference. That dilution factor raises the response threshold of the bioassays to 1.1 μg.mL^{-1} Vg based on the ELISA's sensitivity, or 2.1 μg.mL^{-1} Vg based on the ELISA's PDL.

The interassay variability of the ELISA was assessed by measuring Vg in spiked solutions (25, 80, and 200 ng.mL^{-1}) that corresponded to the low- (20%), medium- (50%), and high- (80%) binding ranges of the calibration curve. The interassay CV was < 10% at the midrange concentration (80 ng.mL^{-1}) but increased for the low (42%) and high (56%) concentrations that correspond to the extremes of the calibration curve where less precision is expected.

ELISA for Rainbow Trout Vg

A calibration curve for the ELISA for rainbow trout Vg is presented in Figure 6.7. The binding in the ELISA was inhibited by 10% at about 12 ng.mL^{-1} Vg; 20% binding inhibition occurred at 19 ng.mL^{-1} Vg. When fish plasma is analyzed in the ELISA at a dilution of 1 in 10, this corresponds to a response threshold of 120–200 ng.mL^{-1} in the final bioassays. The ELISA's CV was < 10% across its working range (Figure 6.7).

Fish Exposures

Immature fish (60–100 g) were used in the bioassays. Fish were exposed to effluents and chemicals by i.p. injection or by waterborne exposure. During experiments the fish were kept in aerated 50-L glass aquaria (5 fish per aquarium) at 8°C for brown trout and 12.5 or 15°C for rainbow trout. The photoperiod was 12 hours.

I.P. Exposure

The i.p.-based bioassay was characterized by following the time course of induction of Vg in brown trout that were treated with a dose series of 17β-estradiol (100-, 500-, and 1000 μg.kg^{-1}; single injection) of 17β-estradiol.[87] Each dose gave a strong response after 11 days (Figure 6.8). At the two higher doses the response appeared to level out between 11 and 15 days, with some evidence of a slight decline following prolonged exposure. The apparently sharp decline in the Vg response for the 100 μg.kg^{-1} dose treatment at 37 days postinjection is derived from a single fish and may not be indicative. No Vg was detected in the negative control fish at any of the time intervals, or at zero time. Based on those results, an exposure period of 10 days was used for experiments in which fish were exposed

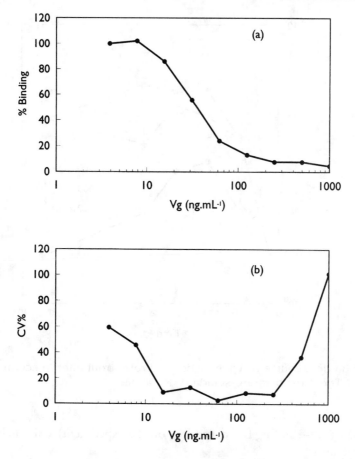

Figure 6.7. Calibration curve (a) and intraassay precision profile (b) for the ELISA for rainbow trout Vg. The precision profile is derived from curve a.

to test solutions by i.p. injection. Donohoe and Curtis[57] exposed rainbow trout to i.p.-administered doses of estradiol, and, although they used a 3-day booster injection, they observed a similar response pattern to the 0.1- and 1-mg.kg^{-1} doses of estradiol to those observed in brown trout in the present study. The time dependence of the response patterns was also similar over 12 days in both studies.

Waterborne Exposure

Test solutions were renewed on alternate days; the renewal volume was either one-third of test volume or 100% except for one pulp mill effluent exposure (50%). A negative control (ethanol 0.1 mL.L^{-1}) and a 17β-estradiol positive con-

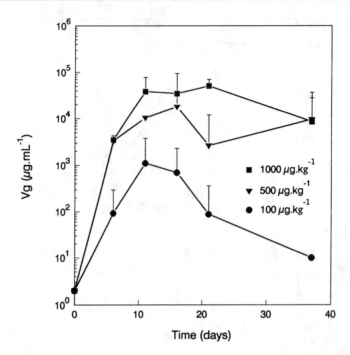

Figure 6.8. Time course of Vg induction in brown trout after injection of 17β-estradiol. The error bars are standard deviations.

trol (100 ng.L^{-1}–1000 ng.L^{-1}, depending on the experiment) were included in each experiment.

Brown Trout

A time course experiment in which brown trout were exposed to a water-borne exposure of 10 μg.L^{-1} 17β-estradiol, which was partially replaced (33%) on alternate days that showed that Vg was strongly induced in the fish after 5 days, and the response seemed to level off after 15 days (data not shown). Based on those data, an exposure period of 15 days was used in the initial exposures of trout to industrial effluents. The concentration dependence of the induction of Vg in trout by 17β-estradiol is presented in Figure 6.9.[88] With the 15-day exposure and 33% renewal protocol, the response appeared to level off at about 1 μg.L^{-1} estradiol. The lowest observed response was at 100 μg.L^{-1}.

To assess the possible influence of the renewal protocol on the response of brown trout to relatively low doses of estradiol, we used a full renewal protocol to examine the time course of Vg induction in brown trout in response to 100

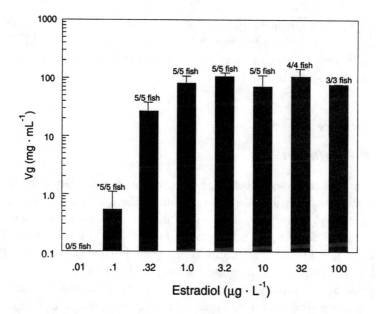

Figure 6.9. Vg induction in brown trout in response to increasing doses of 17β-estradiol. A partial renewal (33%) protocol was used in the 15-day static assay. * Vg induced in 5 out of 5 fish.

μg.L⁻¹ 17β-estradiol—the LOEC (lowest observed effect concentration) from the previous dose response experiment. There was marked Vg induction (1 mg.mL⁻¹) by 10 days, after which the level of induction increased to about 8 mg.mL⁻¹ by 21 days. The plasma Vg concentration in the 21-day exposed fish exceeded that induced in fish that were exposed to the same level of estradiol for 15 days in the partial renewal experiment (0.5 mg.mL⁻¹) by sixteen-fold. For that reason, the effluent in subsequent exposures of brown trout to pulp mill effluent was fully renewed on alternate days for at least 21 days.

Rainbow Trout

Similar optimization experiments were performed for the waterborne bioassay based on rainbow trout. Those results shall be presented elsewhere.

APPLICATION OF BIOASSAYS

Our initial hypothesis was that detectable amounts of EEs are most likely to occur close to point source discharges such as municipal sewage treatment plants, industrial waste water discharges, or sites where intensive animal husbandry is practiced. We are particularly interested in learning whether pulp mill effluents, which are known to contain nonylphenol and phytoestrogens from the wood-furnish, are estrogenic to fish.

I.P. Exposure

Pulp Mill Effluent

The mill selected was a bleached kraft mill with secondary effluent treatment (aerated stabilization basin) located in southern Ontario. The mill used approximately 50% chlorine dioxide substitution. A methanol-extracted pulp and paper mill effluent particulate fraction (RET) was prepared according to Burnison et al.[89] The RET fraction was physically isolated (0.2 μm to 1.0 μm) from the final pulp mill effluent and subsequently freeze-dried and extracted with methanol. Preliminary studies have demonstrated that this fraction significantly induces mixed function oxygenase activity in a static four-day rainbow trout bioassay.[89] The methanol extract was diluted so that 1 mL of the effluent extract was equivalent to 1 L of whole effluent. The effluent extract was further diluted in a solution of methanol in PBS (6:4 [v/v]) and injected into the fish to give the following doses 10-, 100-, and 600-mLequiv.kg^{-1} fish.[90] After the 10-day exposure there was no evidence of Vg induction in the effluent-treated 600-mLequiv.kg^{-1} fish at the ELISA threshold of 1—2.1 μg.mL^{-1}. Vg was induced in the positive-control fish and there was no evidence of Vg induction in fish treated with the carrier alone.

In a second experiment the effluent extract was dissolved in peanut oil : methanol (1:1 [v/v]) and similarly injected into trout to give the following doses: 10-, 100-, and 1,000-mL.kg^{-1}. Booster injections were given on day 5 to compensate for possible metabolic clearance of effluent components. A vehicle control and a positive control (500 μg.kg^{-1} 17β-estradiol) were included in the experiment. After a 10-day exposure there was no evidence of Vg induction in the effluent-treated fish at the ELISA threshold of 1.1—2.1 μg.mL^{-1}. Vg was induced in the positive-control fish and there was no evidence of Vg induction in fish treated with the carrier.

Black Liquor

Black liquor, a recovered by-product in the pulping industry, was obtained from a modernized bleached kraft mill in southern Ontario. Approximately 3% of black liquor is lost to the sewer system which carries the waste to the primary and secondary treatment facilities of the mill before the final effluent is discharged to the receiving waters. An acid precipitate of a 1-mL portion of the black liquor was prepared by the addition of H_2SO_4 until the pH changed from 12.8 to 2.0. The precipitate was extracted with methanol and adjusted to 5 mL. The black liquor extract was further diluted in a solution of methanol in PBS (6:4 [v/v]) and injected i.p. (1 μL.g^{-1}) into trout (5 x) to give the following doses: 10-, 100-, and 375-mL(equiv).kg^{-1}. Dilutions of the black liquor are expressed in units of effluent-equivalents based on the mixed function oxygenase (MFO) inducing potency of black liquor compared to whole effluent: a 0.032% solution of black liquor had an MFO-inducing potency equivalent to 10 L of pulp mill effluent in a static renewal exposure of rainbow trout (data not presented). After a 10-day exposure there was no evidence of Vg induction in the black liquor treated fish at the ELISA threshold of 1.1 to 2.1 μg.mL^{-1}. The Vg responses in the positive- and negative-control fish were normal.

In a second experiment the dilutions of black liquor extract were prepared in a suspension of peanut oil in methanol (1:1) and similarly injected into brown trout to give the following doses 10-, 100-, and 625-mL(equiv).kg^{-1}. The injections were repeated after five days. After a 10-day exposure there was no evidence of Vg induction in the effluent-treated fish at an ELISA threshold of 1.1 to 2.1 μg.mL^{-1}. The Vg responses in the positive- and negative-control fish were again normal.

Zacharewski et al.[81] used an in vitro test system based on MCF-7 cells that were transfected with the human estrogen receptor to detect estrogenic activity in a dilution of the same crude black liquor that was acid precipitated and extracted with methanol for testing in the present study. Negligible activity was detected in a similar pulp mill effluent preparation to the one used in the present study. The transfected MCF-7 cells are extremely sensitive and can detect as little as 5 ng.L^{-1} 17β-estradiol. Such concentrations are likely to lie below the induction threshold of the i.p. bioassays, as only 1 μL.g^{-1} is injected into each fish.

Octylphenol

We also assessed the ability of the known EE OP to induce Vg in brown trout. OP dissolved in ethanol:PBS (6:4 [v:v]) was injected i.p. (1 μL.g^{-1}) into five trout to give a 4-step logarithmic dose series from 3.16 mg.kg^{-1} to 100 mg.kg^{-1}. Vg was induced in the fish treated with 32 and 100 mg.kg^{-1} OP (Table 6.2). Not all fish responded at each dose. There was no evidence of Vg induction in the fish treated with lower doses of OP or in the fish treated with carrier alone at an ELISA threshold level 1.1 to 2.1 μg.mL^{-1}.

Table 6.2. Induction of Vg in Brown Trout by Octylphenol.

Treatment (mg.Kg⁻¹)	Mean Vg (mg.mL⁻¹)
Vehicle control	0
OP[a] 3.2	0
OP 10	0
OP 31.6	2.17
OP 100	4.37
E2[b] 0.5	23.4

[a] OP = octylphenol.
[b] E2 = 17β-estradiol.

Waterborne Exposures

Laboratory Exposure to Pulp Mill Effluents

The ability of pulp mill effluent to induce Vg in brown trout was tested in two experiments.[91] For the first experiment, final effluent was collected from a Kraft softwood mill in Ontario. The mill uses 100% chlorine dioxide substitution and the effluent undergoes secondary treatment in aerated basins. The effluent was brought to the laboratory in 200-L barrels and stored at 4°C. Five fish were exposed to 48-L of a dilution series of effluent (25-, 50-, 75-, and 100%) in a static assay for 15 days (50% renewal on alternate days). After 15-day exposure, there was no evidence of Vg induction in any of the effluent-treated fish or negative-control fish at a threshold of 1.1 to 2.1 µg.mL⁻¹ in the ELISA. Vg was induced to a mean level of 58.8 mg.mL⁻¹ in the plasma of the positive control fish which had been exposed to 17β-estradiol (1 µg.L⁻¹).

For the second experiment, pulp mill effluent was collected from a bleached hardwood kraft mill in Quebec that uses 70% chlorine dioxide substitution and secondary effluent treatment. Five fish were exposed to 100% effluent which was fully replaced every two days for 28 days. There was no evidence of Vg induction in any of the effluent-exposed fish or negative-control fish after a 28-day exposure. Vg was induced to a mean level of 3.9 mg.mL⁻¹ in the plasma of the positive-control fish which had been exposed to 17β-estradiol (100 ng.L⁻¹).

In Situ Exposure to Refinery Effluent

Negative results are often difficult to explain. Our inability to detect Vg induction in fish that were exposed to pulp mill effluents in the laboratory may have several explanations.[91] The effluents might not contain detectable levels of estrogenic chemicals. Estrogens, if present, may not be bioavailable to the fish, or the expression of their activity might have been blocked. Antiestrogens in the effluent could block ER binding sites. Metabolic deactivation, or degradation, could neutralize the potential EE. It is also possible that any estrogens in the effluents are unstable during storage at 4°C. The industrial effluents tested had undergone secondary treatment in biological digestors, which would be likely to remove unstable chemicals. It is also possible that our bioassays are unable to detect estrogens that may be in the effluent, or that the effluents do not contain estrogens. In keeping with the null hypothesis, we assume the latter to be the case until demonstrated otherwise.

One solution to the problem posed by potentially unstable inducers would be to refine the protocols for laboratory-bioassays by the transport of fresh effluent to the laboratory every one or two days. That, however, would be impractical for sites located far from the laboratory. An alternative solution would be to expose fish on site either by means of a flow-through protocol, or by caging fish in the effluent stream, or in the recipient water. The in situ strategy has several advantages: it guarantees a constant supply of the potential inducers; it is less labor-intensive and costly than laboratory assays; and the fish would experience changes in effluent due to process variability.

We caged rainbow trout in a creek that receives effluent from a petroleum refinery.[92] Juvenile fish were held in homemade cages (15 per cage) constructed from plastic laundry baskets that had been previously conditioned by immersion in Great Lakes water for 10 days. The cages were moored for 16 days at several sites above, and one site below, the effluent outfall. No induction of Vg was observed in fish caged at any of the sites. Livers from fish that were caged at a site 20 m above the effluent discharge had elevated MFO (mixed function oxygenase) levels. The levels of induction were significantly ($P < 0.05$) higher than observed in fish caged at a site 30 m below the outfall. The response at the above outfall site may have been caused by billowing of the plume toward the upstream site at the point where the rapidly flowing effluent stream enters the slow-flowing river, followed by dilution with river water as the effluent proceeds to the downstream site. Plume delineation studies are required to verify that suggestion. The detection of MFO activity in the ambient water may have been influenced by the use of a 16-day exposure rather than the more commonly used 3-day protocol. The longer exposure would facilitate the detection of low-level inducers, as well as allowing the detection of induction due to pulse events that would be more likely to be missed during the shorter exposure. Also, the fish were larger than are normally used in our in situ tests for MFO inducers (1–3 g).

We have previously shown that the "end of pipe" effluents from Ontario refineries consistently induce MFO activity in rainbow trout.[93] Thus, the detection of elevated MFO activity in the fish from 20 m above the outfall indicates that those fish were likely exposed to refinery effluent. The failure to detect Vg in the plasma of the exposed fish suggests that the refinery does not discharge high levels of estrogens into the recipient water. Any estrogens in the effluent, if present, were diluted to below the detection threshold of our bioassay by the river water. A longer exposure period for the caged fish, coupled with a flow-through exposure or laboratory test of undiluted effluent, is needed to preclude the possibility that refinery effluent contains low levels of low potency estrogens.

SUMMARY AND CONCLUSIONS

The endocrine system can be disrupted by complex and interacting mechanisms. Recent attention has focused on the ability of xenobiotic or naturally occurring chemicals to mimic the structure and receptor binding properties of endogenous hormones such as the female sex hormone 17β-estradiol. The ability of an exogenous chemical to mimic the receptor binding properties of 17β-estradiol can have profound biochemical, physiological, reproductive, and developmental consequences. The male reproductive system, particularly in its developmental phase, is likely to be susceptible to abnormal estrogen levels. It is important to recognize, however, that the effective levels of estrogen in an organism can be perturbed by mechanisms other than hormone mimicry. For example, an increase in the level or activity of aromatase, which catalyzes the conversion of testosterone to estradiol, could lead to elevated levels of endogenous estrogen and decreased testosterone levels. The impaired metabolism of estradiol could also conceivably lead to unbalanced sex hormone ratios in key organs or tissues. Either of those mechanisms would lead to enhanced production of Vg and a positive response in our bioassays.

Our tests for the detection of environmental estrogens are based on the induction of Vg in brown- and rainbow trout. The tests are intended for the screening of effluents, contaminated waters, chemical mixtures, and pure chemicals for estrogenic effects. The waterborne bioassay can be readily modified for the testing of effluents using on-site flow-through aquaria. Brown trout were used in the bioassays because of their importance as an environmentally sensitive freshwater species that is widely distributed in colder climates. Rainbow trout are a widely used test species that are available in our area at the desired size range (40–100 g) throughout the year.

The working sensitivities of the bioassays based on brown trout (response threshold of 1–2 $\mu g.mL^{-1}$ Vg), which we are in the process of measuring, will be limited primarily by the low-level interference in male plasma, which requires a starting dilution of 1:100, rather than by the sensitivity of the ELISA. The brown trout based bioassays can detect estrogens that induce levels of plasma Vg to greater than 2 $\mu g.mL^{-1}$. The sensitivity and working range of both ELISAs for Vg compare favorably to those from other laboratories (Table 6.1). A more sensitive version of the ELISA for BtVg, however, would allow the detection of lower amounts of Vg at a plasma dilution of 1:100. For example, a detection of 1 ng.mL^{-1} in the ELISA would allow the detection of 100 ng.mL^{-1} of Vg in plasma that is diluted by 1:100. The identity of the cross-reacting molecule in the Bt-ELISA is unknown. A similar effect did not occur when plasma of juvenile rainbow trout maintained on the same feed regime were analyzed by a MAB-based ELISA, which suggests the interference is probably not Vg-induced by phytoestrogens in the feed. Western blot analysis of plasma from male fish should establish whether it is caused by low levels of Vg or another plasma protein.

Based on responses to 17β-estradiol, an 11-day exposure seemed adequate for the i.p.-based bioassay. The ability of octylphenol to induce Vg in the i.p. bioassay shows the assay's responsiveness to an environmental estrogen with a much lower potency than 17β-estradiol.[94] Our initial evaluation of waterborne exposure protocols indicated the combined benefit of increased exposure time (21 days) and full renewal of the test solution in the waterborne tests. Shorter exposure periods and partial renewal protocols could cause the estrogenicity of lower potency solutions to be underestimated. It is also possible that prolonged (>21 days) exposure to even lower doses of estrogens than the 21-day LOEC (lowest observed effect concentration) may also lead to measurable Vg induction. Administration of the test chemical via a flow-through system would be likely to maintain a more consistent concentration of the test chemical, particularly if the chemical is hydrophobic, than either of the static renewal protocols used in the present study.

The present study failed to detect Vg in plasma of trout that were treated with quite high doses of black liquor extract (625 mL [effluent equivalents].mL^{-1}) which corresponds to 0.2 mL of raw black liquor. Zacharewski et al.[81] used an in vitro test system based on recombinant MCF-7 cells to detect estrogenic activity in a dilution of the same crude black liquor. Evidence from the literature suggests that transfected MCF-7 cells are more responsive to estradiol than cultured trout hepatocytes, which suggests that in vivo tests with fish could usefully complement in vitro assays based on transfected MCF-7 cells. A prolonged exposure of trout to waterborne black liquor would help determine whether the low level of inducers detected in the in vitro assay can elicit an estrogenic effect in vivo. The failure of the methanol extract of the particulate fraction of pulp mill effluent to induce Vg, could mean that estrogens in the particulate phase are not extracted by methanol, or that the estrogens are mainly in the effluent's soluble phase. It could also mean that there are no biologically active estrogens in the effluent, or that their potencies, if present, are lower than the threshold response of the Bt-bioassay. Mellanen et al.[95] who treated rainbow trout with 10 mg of freeze-dried pulp mill effluent in silastic implants also failed to observed a Vg response in vivo. In the present study, whole pulp mill effluent also failed to induce an estrogenic effect at a response threshold of 2 µg.mL^{-1} Vg in the waterborne test. Although labile estrogens are likely to be eliminated during the effluent treatment process, the possible loss of estrogenic activity during storage of the effluent at 4 °C can only be discounted through onsite exposures.

Conclusions

Overall, the Vg parameter has several advantages which make it a potentially valuable bioindicator of exposure of fishes to environmental estrogens. The Vg endpoint seems to meet the theoretical prerequisites for a useful bioindicator. Vg induction is a sensitive response parameter. To our knowledge, there are no data in the literature to show that estrogens can cause receptor-mediated effects in fish at doses lower than the Vg induction threshold. In whole fish tests, Vg induction detects both exogenous estrogens and pro-estrogens, which require metabolic activation, while also compensating for metabolic deactivation of exogenous estrogens, if it occurs. Vg induction in whole fish automatically adjusts for the relative potencies of individual chemicals in complex mixtures, while compensating for pharmacokinetic factors and the possible interference of antagonistic chemicals. The Vg endpoint is also likely to detect estrogenic effects caused by mechanisms other than hormone mimicry, such as exogenously induced imbalances in the ratio of endogenous estradiol: testosterone. Vg can be measured precisely, accurately, rapidly, and inexpensively in fish plasma, once antibodies and reagent Vg have been prepared. A practical disadvantage of the Vg endpoint arises from the poor conservation of the protein's immunogenic sites among fish species. For that reason, it is usually necessary to generate anti-Vg for each fish species of interest unless those species are closely related. The versatile Vg endpoint is suited to laboratory bioassays and on-site exposures, whether in situ, using caged fish, or in flow-through chambers. The laboratory bioassays can be based on varied exposure routes including spiked feed, i.p. injections, silastic implants, or waterborne exposures. The properties and availability of the target chemical or mixture will dictate the choice of exposure route. Vg is produced by primary culture of trout hepatocytes upon stimulation with estradiol, which provides the basis for an in vitro assay that can used to screen suspect chemicals, to assess the estrogenicity of mixtures of chemicals, and to measure the estrogenicity of fractionated mixtures, such as industrial wastewaters, in toxicity identification and evaluation (TIE) studies.

We continue to refine our Vg-based bioassays, and are using them to assess the estrogenic potencies of a wide variety of effluents, ambient waters, and suspect chemicals. Together with other approaches the Vg bioindicator can help provide a better understanding of the scale and scope of the environmental estrogen problem in Canada.

ACKNOWLEDGMENTS

We thank Kent Burnison who prepared the pulp mill effluent and black liquor fractions, Anne Borgmann who analyzed many of the samples by ELISA, and Mark Fielden who helped develop the ELISA for brown trout Vg. Portions of this chapter summarize work that has been submitted for publication to *Science of the Total Environment*.

REFERENCES

1. Sumpter, J.P. Feminized responses in fish to environmental estrogens. *Toxicol. Lett.*, 82-3, pp. 737–742, 1995.
2. Guillette, L.J., S.F. Arnold, and J.A. McLachlan. Ecoestrogens and embryos— is there a scientific basis for concern? *Anim. Reprod. Sci.*, 42, pp. 13–24, 1996.
3. Colborn, T., D. Dumanoski, and J.P. Myers. *Our Stolen Future*. Dutton, New York, 1996, p. 306.
4. Safe, S.H. and K. Gaido. Phytoestrogens and anthropogenic estrogenic compounds. *Environ. Toxicol. Chem.*, 17, pp. 119–126, 1998.
5. Safe, S.H. Environmental and dietary estrogens and human health: Is there a problem? *Environ. Health Perspect.*, 103, pp. 346–351, 1995.
6. Hose, J.E. and L.J. Guillette. Defining the role of pollutants in the disruption of reproduction in wildlife. *Environ. Health Perspect.*, 103, pp. 87–91, 1995.
7. Raloff, J. The gender benders. *Sci. News*, 145, pp. 24–27, 1994.
8. Weseloh, D.V., S.M. Teeple, and M. Gilbertson. Double-crested cormorants of the Great Lakes: Egg laying parameters, reproductive failure and contaminant residues in eggs, Lake Huron 1972–1973. *Can. J. Zool.*, 61, pp. 427–436, 1983.
9. Fry, D.M. Reproductive effects in birds exposed to pesticides and industrial chemicals. *Environ. Health Perspect.*, 103, pp. 165–171, 1995.
10. Guillette, L.J., T.S. Gross, G.R. Masson, J.M. Matter, H.F. Percival, and A.R. Woodward. Developmental abnormalities of the gonad and abnormal sex hormone concentrations in juvenile alligators from contaminated and control lakes in Florida. *Environ. Health Perspect.*, 102, pp. 680–688, 1994.
11. Guillette, L.J., D.B. Pickford, D.A. Crain, A.A. Rooney, and H.F. Percival. Reduction in penis size and plasma testosterone concentrations in juvenile alligators living in a contaminated environment. *Gen. Comp. Endocrinol.*, 101, pp. 32–42, 1996.

12. Vonier, P.M., D.A. Crain, J.A. McLachlan, L.J. Guillette, and S.F. Arnold. Interaction of environmental chemicals with the estrogen and progesterone receptors from the oviduct of the American alligator. *Environ. Health Perspect.*, 104, pp. 1318–1322, 1996.

13. Adams, N.R. Detection of the effects of phytoestrogens on sheep and cattle. *J. Anim. Sci.*, 73, pp. 1509–1515, 1995.

14. Knight, D.C. and J.A. Eden. A review of the clinical effects of phytoestrogens. *Obstet. Gynecol.*, 87, Iss 5, Part 2, pp. 897–904, 1996.

15. Giesy J.P., J.P. Ludwig, and D.E. Tillitt. Dioxins, Dibenzofurans, PCBs and Colonial, Fish-Eating Water Birds, in *Dioxins and Health*, Scheckter, A. Ed., Plenum Press, New York, 1994.

16. Giesy, J.P., D.A. Verbrugge, R.A. Othout, W.W. Bowerman, M.A. Mora, P.D. Jones, J.L. Newsted, C. Vandervoort, S.N. Heaton, and S.J. Bursian. Contaminants in fishes from Great Lakes-influenced sections and above dams of 3 Michigan rivers. 2. Implications for health of mink. *Arch. Environ. Contam. Toxicol.*, 27, pp. 213–223, 1994.

17. Walker, M.K. and R.E. Peterson. Aquatic Toxicity of Dioxins and Related Chemicals, in *Dioxins and Health*, Schecter, A. Ed., Plenum Press Div. Plenum Publishing Corp, New York, 1994.

18. Foster, W.G. The reproductive toxicology of Great Lakes contaminants. *Environ. Health Perspect.*, 103, Suppl. 9, pp. 63–69, 1995.

19. Munkittrick, K.R., C.B. Portt, G.J. Van der Kraak, I.R. Smith, and D.A. Rokosh. Impact of bleached Kraft mill effluent on population characteristics, liver MFO activity, and serum steroid levels of a Lake Superior white sucker (*Catostomus commersoni*) population. *Can. J. Fish. Aquat. Sci.*, 48, pp. 1371–1380, 1991.

20. Gagnon, M.M., D. Bussieres, J.J. Dodson, and P.V. Hodson. White sucker (*Catostomus Commersoni*) growth and sexual maturation in pulp mill-contaminated and reference rivers. *Environ. Toxicol. Chem.*, 14, pp. 317–327, 1995.

21. Kovacs, T.G., R.H. Voss, S.R. Megraw, and P.H. Martel. Perspectives on Canadian field studies examining the potential of pulp and paper mill effluent to affect fish reproduction. *J. Toxicol. Environ. Health*, 51, pp. 305–352, 1997.

22. Bortone, S.A. and W.P. Davis. Fish intersexuality as indicator of environmental stress. *BioScience*, 44, pp. 165–172, 1994.

23. Cody, R.P. and S.A. Bortone. Masculinization of mosquitofish as an indicator of exposure to Kraft mill effluent. *Bull. Environ. Contam. Toxicol.*, 58, pp. 429–436, 1997.

24. Purdom, C.E., P.A. Hardiman, V.J. Bye, N.C. Eno, C.R. Tyler, and J.P. Sumpter. Estrogenic effects of effluents from sewage treatment works. *Chem. Ecol.*, 8, pp. 275–285, 1994.

25. Lazier, C.B. and M.E. Mackay. Vitellogenin Gene Expression in Teleost Fish, in *Biochemistry and Molecular Biology of Fishes*, Hochachka, P.W. and T.P. Mommsen, Eds., Elsevier Science Publishers, B.V., 1993.

26. Specker, J.L. and C.V. Sullivan. Vitellogenesis in fishes: Status and perspectives, in *Perspectives in Comparative Endocrinology*. Davey, K.G., R.E. Peter, and S.S. Tobe, Eds., National Research Council of Canada, Ottawa, 1994.

27. Boockfor, F.R. and C.A. Blake. Chronic administration of the environmental pollutant 4-tert-octylphenol to adult male rats interferes with the secretion of luteinizing hormone, follicle-stimulating hormone, prolactin, and testosterone. *Biol. Reprod.*, 57, pp. 255–266, 1997.

28. Sumpter, J.P. and S. Jobling. Vitellogenesis as a biomarker for estrogenic contamination of the aquatic environment. *Environ. Health Perspect.*, 103, pp. 173–178, 1995.

29. de Valming, V.L., H.S. Wiley, G. Delahunty, and R.A. Wallace. Goldfish (*Carassius auratus*) vitellogenin: Induction, isolation, properties, and relationship to yolk proteins. *Comp. Biochem. Physiol.*, 67B, pp. 613–623, 1980.

30. Norberg, B. and C. Haux. Induction, isolation and a characterization of the lipid content of plasma vitellogenin from two Salmo species: Rainbow trout *(Salmo gairdneri)* and sea trout *(Salmo trutta)*. *Comp. Biochem. Physiol.*, 81B, pp. 869–876, 1985.

31. Sumpter, J.P. The Purification, Radioimmunoassay and Plasma Levels of Vitellogenin from Rainbow Trout, (*Salmo gairdneri*), in *Current Trends in Comparative Endocrinology*, Lofts, B. and W.N. Holmes, Eds., Hong Kong University Press, Hong Kong, 1985.

32. Rodriguez, J.N., O. Kah, M. Geffard, and F.L. Menn. Enzyme-linked immunosorbent assay (ELISA) for sole (*Solea vulgaris*) vitellogenin. *Comp. Biochem. Physiol.*, 92B, pp. 741–746, 1989.

33. Kishida, M., T.R. Anderson, and J.L. Specker. Induction by beta-estradiol of vitellogenin in striped bass (*Morone saxatilis*): Characterization and quantification in plasma and mucus. *Gen. Comp. Endocrinol.*, 88, pp. 29–39, 1992.

34. Gordon, M.R., T.G. Owen, T.A. Ternan, and L.D. Hildebrand. Measurement of a sex-specific protein in skin mucus of premature coho salmon (*Oncorhynchus kisutch*). *Aquaculture*, 43, pp. 333–339, 1984.

35. Valotaire, Y., M.G. Leroux, and P. Jego. Estrogen receptor gene-structure and expression in rainbow trout. *Mol. Biol. Frontiers*. 2, pp. 373–390, 1993.

36. Smith, C.L., O.M. Conneely, and B.W. Omalley. Oestrogen receptor activation in the absence of ligand. *Biochem. Soc. Trans.*, 23, pp. 935–939, 1995.

37. Thomas, P. Effects of Aroclor 1254 and cadmium on reproductive endocrine function and ovarian growth in Atlantic Croaker. *Mar. Environ. Res.*, 28, pp. 499–503, 1989.

38. Tam, W.H., L. Birkett, R. Makaran, P.D. Payson, D.K. Whitney, and Ck.-C. Yu. Modification of carbohydrate metabolism and liver vitellogenic function in brook trout (*Salvelinus fontinalis*) by exposure to low pH. *Can. J. Fish. Aquat. Sci.*, 44, pp. 630–635, 1987.

39. Parker, D.B. and B.A. Mckeown. Effects of pH and/or calcium-enriched freshwater on plasma levels of vitellogenin and Ca^{2+} and on bone calcium

content during exogenous vitellogenesis in rainbow trout *Salmo gairdneri*. *Comp. Biochem. Physiol.*, 87A, pp. 267–273, 1987.

40. Mackay, M.E. and C.B. Lazier. Estrogen responsiveness of vitellogenin gene expression in rainbow trout *(Oncorhynchus mykiss)* kept at different temperatures. *Gen. Comp. Endocrinol.* 89, pp. 255–266, 1993.

41. Chen, T.T., P.C. Reid, R.V. Beneden, and R.A. Sonstegard. Effect of Aroclor 1254 and mirex on estradiol-induced vitellogenin production in juvenile rainbow trout *(Salmo gairdneri)*. *Can. J. Fish. Aquat. Sci.*, 43, pp. 169–173, 1986.

42. Harries, J.E., D.A. Sheahan, S. Jobling, P. Matthiessen, P. Neall, E.J. Routledge, R. Rycroft, J.P. Sumpter, and T. Taylor. A survey of estrogenic activity in United Kingdom inland waters. *Environ. Toxicol. Chem.*, 15, pp. 1993–2002, 1996.

43. Harries, J.E., D.A. Sheahan, S. Jobling, P. Matthiessen, M. Neall, J.P. Sumpter, T. Taylor, and N. Zaman. Estrogenic activity in five United Kingdom rivers detected by measurement of vitellogenesis in caged male trout. *Environ. Toxicol. Chem.*, 16, pp. 534–542, 1997.

44. Lech, J.J., S.K. Lewis, and L. Ren. *In vivo* estrogenic activity of nonylphenol in rainbow trout. *Fundam. Appl. Toxicol.*, 30, pp. 229–232, 1996.

45. White, R., S. Jobling, S.A. Hoare, J.P. Sumpter, and M.G. Parker. Environmentally persistent alkylphenolic compounds are estrogenic. *Endocrinology*, 135, pp. 175–182, 1994.

46. Lye, C.M., C.L.J. Frid, M.E. Gill, and D. McCormick. Abnormalities in the reproductive health of flounder *Platichthys flesus* exposed to effluent from a sewage treatment works. *Mar. Pollut. Bull.*, 34, pp. 34–41, 1997.

47. Katzenellenbogen, J.A. The structural pervasiveness of estrogenic activity. *Environ. Health Perspect.*, 103, pp. 99–101, 1995.

48. Soto, A.M., H. Justicia, J.W. Wray, and C. Sonnenschein. *p*-Nonyl-phenol: An estrogenic xenobiotic released from "modified" polystyrene. *Environ. Health Perspect.*, 92, pp. 167–173, 1991.

49. Shelby, M.D., R.R. Newbold, D.B. Tully, K. Chae, and V.L. Davis. Assessing environmental chemicals for estrogenicity using a combination of *in vitro* and *in vivo* assays. *Environ. Health Perspect.*, 104, pp. 1296–1300, 1996.

50. Flouriot, G., F. Pakdel, B. Ducouret, and Y. Valotaire. Influence of xenobiotics on rainbow trout liver estrogen receptor and vitellogenin gene expression. *J. Mol. Endocrinol.* 15, pp. 143–151, 1995.

51. Routledge, E.J. and J.P. Sumpter. Estrogenic activity of surfactants and some of their degradation products assessed using a recombinant yeast screen. *Environ. Toxicol. Chem.*, 15, pp. 241–248, 1996.

52. Jobling, S., D. Sheahan, J.A. Osborne, P. Matthiessen, and J.P. Sumpter. Inhibition of testicular growth in rainbow trout *(Oncorhynchus mykiss)* exposed to estrogenic alkylphenolic chemicals. *Environ. Toxicol. Chem.*, 15, pp. 194–202, 1996.

53. Gray, M.A. and C.D. Metcalfe. Induction of testis-ova in Japanese medaka (*Oryzias latipes*) exposed to p-nonylphenol. *Environ. Toxicol. Chem.*, 16, pp. 1082–1086, 1997.

54. Blackburn, M.A. and M.J. Waldock. Concentrations of alkylphenols in rivers and estuaries in England and Wales. *Wat. Res.* 29, pp. 1623–1629, 1995.

55. Field, J.A. and R.L. Reed. Nonylphenol polyethoxy carboxylate metabolites of nonionic surfactants in U.S. paper mill effluents, municipal sewage treatment plant effluents, and river waters. *Environ. Sci. Technol.*, 30, pp. 3544–3550, 1996.

56. Chakravorty, S., B. Lal, and T.P. Singh. Effect of endosulfan (*thiodan*) on vitellogenesis and its modulation by different hormones in the vitellogenic catfish *Clarias batrachus*. *Toxicology*, 75, pp. 191–198, 1992.

57. Donohoe, R.M. and L.R. Curtis. Estrogenic activity of chlordecone, o,p'-DDT and o,p'-DDE in juvenile rainbow trout: induction of vitellogenesis and interaction with hepatic estrogen binding sites. *Aquat. Toxicol.*, 36, pp. 31–52, 1996.

58. McKinney, J.D. and C.L. Waller. Polychlorinated biphenyls as hormonally active structural analogues. *Environ. Health Perspect.*, 102, pp. 290–297, 1994.

59. Gierthy, J.F., K.F. Arcaro, and M. Floyd. Assessment of PCB estrogenicity in a human breast cancer cell line. *Chemosphere*, 34, pp. 1495–1505, 1997.

60. Korach, K.S., P.J. Sarver, K. Chae, J.A. McLachlan, and J.D. McKinney. Estrogen receptor binding activity of polychlorinated hydroxybiphenyls: conformationally restricted structural probes. *Mol. Pharmacol.*, 33, pp. 120–126, 1988.

61. Pelissero, C., B. Bennetau, P. Babin, F. Le-Menn, and J. Dunogues. The estrogenic activity of certain phytoestrogens in the Siberian sturgeon *Acipenser baeri*. *J. Steroid Biochem. Mol. Biol.*, 38, pp. 293–299, 1991.

62. Herman, R.L. and H.L. Kincaid. Pathological effects of orally administered estradiol to rainbow trout. *Aquaculture*, 72, pp. 165–172, 1988.

63. Pereira, J.J., J. Ziskowski, R. Mercaldo-Allen, C. Kuropat, D. Luedke, and E. Gould. Vitellogenin in winter flounder (*Pleuronectes americanus*) from Long Island Sound and Boston Harbor. *Estuaries*, 15, pp. 289–297, 1992.

64. Thomas, P. and J. Smith. Binding of xenobiotics to the estrogen receptor of spotted seatrout: A screening assay for potential estrogenic effects. *Mar. Environ. Res.*, 35, pp. 147–151, 1993.

65. Benefey, T.J., E.M. Donaldson, and T.G. Owen. An homologous radioimmunoassay for Coho salmon (*Oncorhynchus kisutch*) vitellogenin, with general applicability to other pacific salmonids. *Gen. Comp. Endocrinol.*, 75, pp. 78–82, 1989.

66. Copeland, P.A., J.P. Sumpter, T.K. Walker, and M. Croft. Vitellogenin levels in male and female rainbow trout (*Salmo gairdneri* Richardson) at various stages of the reproductive cycle. *Comp. Biochem. Physiol.*, 83, pp. 487–493, 1986.

67. Heppell, S.A., N.D. Denslow, L.C. Folmar, and C.V. Sullivan. Universal assay of Vitellogenin as a biomarker for environmental estrogens. *Environ. Health Perspect.*, 103, pp. 9–15, 1995.

68. Mourot, B. and B.P.Y. Le. Enzyme-linked immunosorbent assay (ELISA) for rainbow trout (*Oncorhynchus mykiss*) vitellogenin. *J. Immunoassay*, 16, pp. 365–377, 1995.

69. Kwon, H.-C., A. Hara, Y. Mugiya, and J. Yamada. Enzyme linked-immunosorbent assay (ELISA) of vitellogenin in whitespotted charr, *Salvelinus leucomaenis. Bull. Fac. Fish. Hokkaido Univ.*, 41, pp. 162–180, 1990.

70. Cuisset, B., C. Pelissero, F. Le-Menn, and J. Nunez-Rodriguez. ELISA for Siberian Sturgeon (*Acipenser baeri* Brandt) Vitellogenin, in *Proceedings of the First International Symposium on Sturgeon, Bordeaux Gironde, France, 3–6 October 1989.* Williot, P., Ed., 1991.

71. Tao, Y., A. Hara, R.G. Hodson, L.C. Woods III, and C.V. Sullivan. Purification, characterization and immunoassay of striped bass (*Morone saxatilis*) vitellogenin. *Fish Physiol. Biochem.*, 12, pp. 31–46, 1993.

72. So, Y.P., D.R. Idler, and S.J. Hwang. Plasma vitellogenin in landlocked Atlantic salmon (*Salmo salar* Ouananiche): Isolation, homologous radioimmunoassay and immunological cross-reactivity with vitellogenin from other teleosts. *Comp. Biochem. Physiol.*, 81B, pp. 63–71, 1985.

73. Norberg, B. and C. Haux. An homologous radioimmunoassay for brown trout (*Salmo trutta*) vitellogenin. *Fish Physiol. Biochem.*, 5, pp. 59–68, 1988.

74. Tyler, C.R. and J.P. Sumpter. The development of a radioimmunoassay for carp, *Cyprinus carpio*, vitellogenin. *Fish Physiol. Biochem.*, 8, pp. 129–140, 1990.

75. Goodwin, A.E., J.M. Grizzle, J.T. Bradley, and B.H. Estridge. Monoclonal antibody-based immunoassay of vitellogenin in the blood of male channel catfish (*Ictalurus punctatus*). *Comp. Biochem. Physiol.*, 101B, pp. 441–446, 1992.

76. Copeland, P.A. and P. Thomas. The measurement of plasma vitellogenin levels in a marine teleost, the spotted seatrout (*Cynoscion nebulosus*) by homologous radioimmunoassay. *Comp. Biochem. Physiol.*, 91B, pp. 17–23, 1988.

77. Mananos, E., J. Nunez, S. Zanuy, M. Carrillo, and F. Lemenn. Sea bass (*Dicentrarchus labrax* L.) vitellogenin. II. Validation of an enzyme-linked immunosorbent assay (ELISA). *Comp. Biochem. Physiol.*, 107, pp. 217–223, 1994.

78. Okumura, H., A. Hara, F. Saeki, T. Todo, S. Adachi, and K. Yamauchi. Development of a sensitive sandwich enzyme-linked immunosorbent assay (ELISA) for vitellogenin in the Japanese eel *Anguilla japonica. Fish. Sci.*, 61, pp. 283–289, 1995.

79. Tyler, C.R., B. van der Eerden, S. Jobling, G. Panter, and J.P. Sumpter. Measurement of vitellogenin, a biomarker for exposure to oestrogenic chemi-

cals, in a wide variety of cyprinid fish. *J. Comp. Physiol. B - Biochem. System Environ. Physiol.*, 166, pp. 418–426, 1996.

80. Matsubara, T., T. Wada, and A. Hara. Purification and establishment of ELISA for vitellogenin of Japanese sardine (*Sardinops melanostictus*). *Comp. Biochem. Physiol. B – Biochem. Mol. Biol.*, 109, pp. 545–555, 1994.

81. Zacharewski, T.R., K. Berhane, B.E. Gillesby, and B.K. Burnison. Detection of estrogen- and dioxin-like activity in pulp and paper mill black liquor and effluent using *in vitro* recombinant receptor reporter gene assays. *Environ. Sci. Technol.*, 29, pp. 2140–2146, 1995.

82. Le Drean, Y., L. Kern, F. Pakdel, and Y. Valotaire. Rainbow trout estrogen receptor presents an equal specificity but a differential sensitivity for estrogens than human estrogen receptor. *Mol. Cell. Endocrinol.*, 109, pp. 27–35, 1995.

83. Soto, A.M., K.L. Chung, and C. Sonnenschein. The pesticides endosulfan, toxaphene, and dieldrin have estrogenic effects on human estrogen-sensitive cells. *Environ. Health Perspect.*, 102, pp. 380–383, 1994.

84. Anderson, M.J., M.R. Miller, and D.E. Hinton. *In vitro* modulation of 17beta-estradiol-induced vitellogenin synthesis: effects of cytochrome P4501A1 inducing compounds on rainbow trout (*Oncorhynchus mykiss*) liver cells. *Aquat. Toxicol.*, 34, pp. 327–350, 1996.

85. Wiley, H.S., L. Opresko, and R.A. Wallace. New methods for the purification of vertebrate vitellogenin. *Anal. Biochem.*, 97, pp. 145–152, 1979.

86. Ekins, R.P. The Precision Profile: Its Use in Assay Design, Assessment, and Quality Control, in *Immunoassays for Clinical Chemistry*, Hunter, W.M. and J.E.T. Corrie, Eds., Churchill-Livingstone, Edinburgh, 1983.

87. Gamble, A., J. Sherry, M. Fielden, P. Hodson, and K. Solomon. Enzyme Linked Immunosorbent Assay (ELISA) for Brown Trout Vitellogenin for Use in the Detection of Environmental Estrogens, Presented at the 2nd SETAC World Congress, Nov. 5–9, Vancouver, B.C., 1995.

88. Gamble, A., J. Sherry, P. Hodson, J. Parrott, and K. Solomon. Bioassay for the Detection of Environmental Estrogens, Presented at 17th Annual Meeting SETAC, Washington, 17–21 Nov., 1996.

89. Burnison, B.K., P.V. Hodson, D.J. Nuttley, and S. Efler. A b.leached-kraft mill effluent fraction causing induction of a fish mixed-function oxygenase enzyme. *Environ. Toxicol. Chem.*, 15, pp. 1524–1531, 1996.

90. Sherry, J., A. Gamble, B. Burnison, P. Hodson, J. Parrott, and K. Solomon. Bioassay of Estrogenic Activity: Vitellogenin Induction in Whole Trout and in Cultured Hepatocytes. Presented at 17th Annual Meeting SETAC, Washington, 17–21 Nov., 1996.

91. Gamble, A., J. Sherry, P. Hodson, K. Solomon, P. Hansen, B. Hock, and A. Marx. Use of Fish Bioassays to Test Effluents for Estrogenic Effects, Presented at 18th Annual Meeting SETAC, San Francisco, 16–20 Nov., 1997.

92. Gamble, A., J. Sherry, P. Hodson, and K. Solomon. *In Situ* Exposures for the Detection of Environmental Estrogens, Presented at 18th Annual Meeting SETAC, San Francisco, 16–20 Nov., 1997.

93. Sherry, J.P., B.F. Scott, J. Parrott, P. Hodson, and S. Rao. The Sublethal Effects of Petroleum Refinery Effluents: Mixed Function Oxygenase (MFO) Induction in Rainbow Trout, Presented at 4th International Conference on Aquatic Ecosystem Health, Coimbra, Portugal, May 14–18, 1995.

94. Jobling, S. and J.P. Sumpter. Detergent components in sewage effluent are weakly oestrogenic to fish: An in vitro study using rainbow trout (*Oncorhynchus mykiss*) hepatocytes. *Aquat. Toxicol.,* 27, pp. 361–372, 1993.

95. Mellanen, P., T. Petanen, J. Lehtimaki, S. Makela, G. Bylund, B. Holmbom, E. Mannila, A. Oikari, and R. Santti. Wood-derived estrogens: Studies *in vitro* with breast cancer cells and *in vivo* in trout. *Toxicol. Appl. Pharmacol.,* 136, pp. 381–388, 1996.

96. Redding, M.J. and R. Patino. Reproductive Physiology, in *The Physiology of Fishes,* Evans, D.H., Ed., 1993, CRC Press, Boca Raton, FL.

97. Bennie, D.T., C.A. Sullivan, H.-B. Lee, T.E. Peart, and R.J. Maguire. Occurrence of alkylphenols and alkylphenol mono- and diethoxylates in natural waters of the Laurentian Great Lakes Basin and the Upper St. Lawrence River. *Sci. Total Environ.,* 193, pp. 263–275, 1997.

98. Zava, D.T., M. Blen, and G. Duwe. Estrogenic activity of natural and synthetic estrogens in human breast cancer cells in culture. *Environ. Health Perspect.,* 105, Suppl. 3, pp. 637–645, 1997.

Chapter 7

Ecotoxicological Assessment of Japanese Industrial Effluents Using a Battery of Small-Scale Toxicity Tests

T. Kusui and C. Blaise

INTRODUCTION

Direct entrance of pollutants into aquatic ecosystems via complex liquid industrial waste emissions continues to be an important area of concern because of the potentially serious consequences which ecotoxic inputs can have on receiving water biota.[1-3] An obvious first step aimed at reducing or eliminating problematic point source discharges requires assessment of their hazard potential and the identification of putative chemicals linked to any observed toxic effects. Several combined biological (toxic effects identification using bioassays) and chemical (contaminant links to observed effects) strategies have been proposed and applied for this purpose since the mid-1980s.[4-7] Within such strategies are those that take advantage of small-scale biotest batteries, employing (micro)organisms representative of different biological levels, because of the attractive features of simplicity, sensitivity, and cost-efficiency which they tend to offer.[8-11]

The opportunity to assess the ecotoxic potential of wastewater samples from industrial and sewage plants located in Toyama Prefecture, Japan, came under a Canada/Japan bilateral project struck between the present authors in 1995. In the study reported herein, 20 composite samples taken from 18 industrial facilities and 2 sewage treatment plants were evaluated with a suite of small-scale bioassays comprising a mixture of both well-standardized and promising new tests. The objective of our cooperative work attempted: (1) to compare the relative performance and sensitivity responses of the bioassays in detecting and quantifying effluent toxicity, (2) to identify physicochemical characteristics analyzed in the samples that could be potentially linked to bioassay responses, and (3) to compare the toxic loadings of these 20 Japanese-based effluents among themselves as well as in relation to loadings of Canadian-based effluents previously-determined under the first Saint-Lawrence River Action Plan.[12]

MATERIALS AND METHODS

Effluent Samples

Eighteen representative industrial sites and two sewage treatment plants from Toyama Prefecture in Japan (Figure 7.1) were visited between October 23 and December 18, 1995, thanks to previous arrangements made with the first author of this study and industry officials. At each site, 4-L grab samples of final effluent were collected in glass vessels and brought back to the laboratory within 4 hours in coolers kept below 10°C with ice packs. At facilities where chlorine treatment was employed, samples were collected before this step was applied since our main goal was to assess toxicants other than residual chlorine. The types of plants investigated and the physicochemical characteristics measured in their effluents, including flow rate information, are shown in Table 7.1.

In preparation for chemical analyses and bioassays, samples were subjected to a three-step filtration (with a 1.0 μ glass fiber filter, followed by 0.45 μ and then 0.22 μ filtration with cellulose membrane filters). Electrical conductivity (EC), OD_{260}, and pH were measured in the 0.45 μ filtrates, while dissolved organic carbon (DOC) and metal analyses as well as bioassay experiments were performed in the 0.22 μ filtrates. Hence, bioassay results presented herein exclusively reflect toxicity present in the soluble fraction of these wastewaters. After filtration, samples were stored at 4°C in glass bottles and testing was initiated within 4 days of collection dates. Heavy metal analyses were carried out by atomic absorption according to standard practice.[13]

Bioassays

The suite of five bioassays chosen for this study represented three trophic levels (decomposers, primary producers and primary/secondary consumers) as well as two ecotoxicity levels (lethality and sublethality). They were selected on the basis of practical criteria including sensitivity, simplicity in undertaking assays, ease in laboratory maintenance of organisms (in the case of algae and Hydra), cost-effectiveness, reliability of experimental procedure, and/or frequency of use (inter)nationally. Test principles for each bioassay are briefly recalled below.

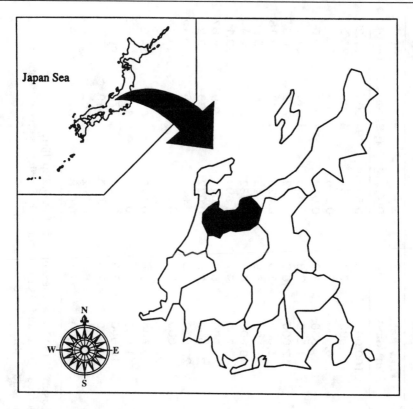

Figure 7.1. Toyama prefecture in Japan where investigated industrial sites and sewage treatment plants are located.

The Microtox® test was employed to report toxicity at the bacterial level. This microassay makes use of a lyophilized luminescent bacterial reagent (*Vibrio fischeri*) which, following acute exposure to liquid samples containing toxic activity, will display a reduction in its light output. Since its initial marketing in 1979,[14] this low-volume-requiring test has proved to be an effective tool to screen varied samples for the presence of bioavailable (in)organic toxicants. Microtox® testing was undertaken with the Environment Canada standard procedure.[15] Bacterial light-production was monitored after exposure times of 5-, 15-, and 30-minutes to effluent samples, but 15-minutes exposure endpoint values were used to report and compare toxicity with that of other bioassays.

Phytotoxicity testing was conducted with the green microalga *Selenastrum capricornutum* (ATCC 22662). The microplate assay procedure initially developed by Blaise et al.[16] was used to report algal growth inhibition after a 72-hour exposure to effluent samples by following an optimized Environment Canada

Table 7.1. Characteristics of Final Effluents Used in this Study.[a]

Effluent	Process	Treatment[b]	Flow Rate (m³/day)	pH	OD$_{260}$	EC[c] (µS/cm)	DOC[d] (mg/L)
1	Pulp and paper	2ʸ + PC	100,000	6.6	0.572	294,000	71.2
2	Chemical production	2ʸ	24,000	7.2	0.357	4,690	23.3
3	Pharmaceuticals	2ʸ	16,000	6.7	0.586	2,700	32.7
4	Surface processing	PC	850	7.0	0.042	1,000	10.0
5	Pharmaceuticals	2ʸ	1,200	7.0	0.438	1,340	37.0
6	Recycled paper	2ʸ	17,000	6.8	0.342	772	26.3
7	Metal processing	PC	3,000	6.3	0.020	1,120	2.18
8	Meat processing	2ʸ	600	6.9	0.112	786	5.96
9	Sewage treatment	2ʸ	49,600	6.5	0.082	280	4.4
10	Textiles	PC	700	9.7	2.994	858	79.1
11	Recycled pulp	PC	7,000	6.9	0.044	399	5.87
12	Chemical production	PC	24,000	7.0	0.022	529	2.2
13	Aluminium production	PC	22,830	7.3	0.002	376	3.45
14	Semiconductor chip production	2ʸ + PC	5,000	8.1	0.011	220	1.73
15	Chemical production	2ʸ	2,400	7.4	0.078	458	34.2
16	Aluminium production	PC	1,000	7.2	0.021	430	6.82
17	Recycled paper	PC	10,000	7.9	0.101	371	60.1
18	Petrochemical refinery	PC	1,600	7.5	0.023	419	3.32
19	Laundry cleaning facility	2ʸ	200	8.9	0.092	762	7.08
20	Sewage treatment	2ʸ	40,000	7.1	0.078	315	4.43

[a] After 0.45µ filtration, pH, OD260, and EC were measured. DOC was measured in the 0.22µ filtrate.

[b] 2ʸ: effluent with both primary and secondary (biological) treatment. PC: effluent with physicochemical treatment including neutralization, flocculation, sand filtration and activated carbon treatment.

[c] Electrical conductivity.

[d] Dissolved organic carbon.

standard protocol.[17] This simple and practical assay has been shown to be valuable in appraising the toxic potential of effluents and various (in)organic compounds.[18]

Microinvertebrate testing representing the primary consumer group was carried out with a 48-hour exposure *Daphnia magna* motility inhibition assay and with the cyst-based *Thamnocephalus platyurus* (Thamnotoxkit) 24-hours acute exposure lethality assay. The first of these has now become an internationally-recognized sensitive assay of long standing for toxicity screening since its initial development.[19] Experimentation followed the standard methodology proposed by Environment Canada.[20] The second assay is one of a series of newly commercialized cyst-based maintenance-free microbiotests, sold as Toxkits, and easily undertaken in specially designed multiwell plates (Creasel, Deinze, Belgium). Subsequent to a 24-hour hatching period, neonates of *T. platyurus* are ready to be inoculated into a multiwell plate containing serial dilutions of liquid samples under study.[21] After a 24-hour exposure to a liquid sample, motionless larvae are rapidly recorded as dead organisms. Because of its operational simplicity and sensitivity, this new assay holds promise for rapid and cost-effective toxicity screening of diverse liquid matrices.

Finally, microinvertebrate testing representing the secondary consumer group was performed with the freshwater cnidarian *Hydra attenuata*. This is a robust animal capable of withstanding extensive manipulation that can be easily reared and maintained in the laboratory. *Hydra* undergo progressive and marked morphological changes upon toxic exposure, allowing for unambiguous recording of concurrent sublethal and lethal effects during testing. We recently developed and described a 96-hour exposure microplate-based procedure to conduct acute toxicity screening of industrial effluents with this organism.[22] Besides demonstrating adequate sensitivity in detecting effluent toxicity, this assay has proved useful in recent toxicity screening of freshwater sediment interstitial waters,[23] as well as in the appraisal of heavy metal toxicity (unpublished results).

Data Analysis

To get an idea of overall bioassay sensitivity, qualitative responses were first estimated by recording the presence or absence of a toxic effect at the highest effluent concentration tested. It varied according to bioassay protocol: 100% v/v for *Hydra*, *Daphnia*, and *Thamnocephalus;* 91% v/v for *Selenastrum;* 45% v/v for Microtox. When concentration-response curves could be generated, 50% endpoint values and associated 95% confidence intervals were determined from linear regression data. Again, according to bioassay protocol, 96-hours LC50s and EC50s were determined for *Hydra*, 48-hours EC50s for *Daphnia*, 24-hours LC50s for *Thamnocephalus,* 72-hours IC50s for *Selenastrum,* and 15-minutes IC50s for Microtox.

For different purposes described later on, 50% effect endpoint values were also transformed into toxic units (TU), according to a concept proposed by Sprague and Ramsay,[24] with the following formula:

TU = 100% v/v *divided by* the 50% effect endpoint value for each bioassay (expressed in % v/v)

In order to evaluate the (dis)similarities of bioassay responses in reporting effluent toxicity, a Pearson product-moment correlation coefficient matrix[25] was determined from TUs generated with each bioassay for the 20 effluents. Spearman rank order correlations were also undertaken between endpoint values and effluent physicochemical characteristics (pH, conductivity, DOC, OD_{260}) in an attempt to explain bioassay toxicity responses.

When examining ways to eliminate the harmful effects of effluents, it is important to determine the cause(s) responsible for toxicity responses in bioassays whenever complementary chemical information is available. In this work, concentrations for six heavy metals were determined in the effluents studied. Overall, minimum to maximum concentrations reported (in $\mu g/L$) were, respectively, 0.5–1.1 for Cd, 11–129 for Cr, 2–320 for Cu, 17.5–3,180 for Mn, 21–1,950 for Ni, and 70–4,440 for Zn. In an attempt to identify the putative metals contributing to the toxicity responses for each of the six bioassays, we calculated a toxicity factor (TF) for each heavy metal as follows:

$$TF_i^m = \frac{C_i^m}{EC_{50} \text{ or } LC_{50} \text{ or } IC_{50}}$$

where: TF_i^m is the toxicity factor of heavy metal *i* in effluent *m*
C_i^m is the measured concentration (mg/L) of heavy metal in effluent *m*
EC_{50}, LC_{50}, and IC_{50} are the endpoint values (mg/L) of heavy metal *i* for each bioassay

According to this concept, the higher the TF_i^m, the greater the chance that heavy metal *i* contributes a toxic effect. If we assume that no interactions occur among metals, we can integrate the six heavy metal TFs in effluent *m* as:

$$TF^m = \sum_1^6 TF_i^m$$

where TF^m represents the overall toxic potential that the six heavy metals can exert in effluent m. Afterward, we can then estimate the combined (or relative singular) effects of the heavy metals on an effluent's toxicity by comparing TF^ms with TUs generated by each bioassay.

Finally, to estimate the relative toxic contribution of each effluent to the receiving environment on a common comparative basis, toxic loadings and PEEP (« Potential Ecotoxic Effects Probe ») values were calculated with the help of the following formula:[10]

$$P = \log_{10}\left[1 + n\left(\frac{\sum_{i=1}^{N} T_i}{N} \right) Q \right]$$

where: P = PEEP numerical value
 n = # of biotests exhibiting a (geno)toxic response
 N = maximum # of *measurable* responses
 T_i = toxic units generated by each biotest before *and* after biodegradability testing of the effluent sample
 Q = effluent flow (m^3/h)

Here, more sensitive 20% endpoint effect values (i.e., LC20s, EC20s, and IC20s), which approximate threshold values determined from NOECs and LOECs in the Canadian study,[10] were transformed into TUs and integrated into the PEEP formula. This enabled further (and closer) comparison of the toxic potential of the Japanese-studied effluents with that of the Canadian effluents for which PEEP values were determined under the first Saint-Lawrence Action Plan from 1988–1993.

RESULTS AND DISCUSSION

Qualitative and quantitative toxicity responses generated by each bioassay toward the effluents are reported in Table 7.2. By collating these values, a relative idea of bioassay sensitivity in detecting and quantifying effluent toxicity can be obtained (Table 7.3). When tested at its highest concentration (45% v/v), the Microtox test, for example, showed light inhibition to 12 of the 20 effluent samples (Table 7.2) indicating a combined 60% response to the 20 effluents (Table 7.3). Its quantitative (concentration-response) capacity to determine discrete IC_{50} values was observed to be 25% (Table 3). The most intense toxic response came from effluent 12 (15-min IC_{50} = 5.9% v/v or 16.9 TU). The range of TUs mea-

Table 7.2. Qualitative and Quantitative Toxicity Responses of Bioassays.

Effluent	Presence (+) or Absence (−) of Toxicity at Response Highest Test Concentration[a]					50% Endpoint Effect Concentration[b]				
	M	D	T	H[c]	S	M	D	T	H	S
1	+	+	−	+	+	—	>100^d	—	68.9	9.8
2	+	+	+	+	+	>100^d	60.2	72.0	15.0	40.0
3	+	+	+	+	+	92.5	51.8	37.5	16.3	13.5
4	+	+	+	+	+	100	9.6	46.2	75.0	43.0
5	+	−	−	−	+	>100^d	—	—	—	>100^d
6	−	−	+	−	+	—	—	—	—	59.6
7	+	+	+	+	+	10.2	4.1	12.9	18.8	2.7
8	−	−	−	+	+	—	—	—	76.5	7.2
9	−	−	−	−	−	—	—	—	—	>100^d
10	+	+	+	+	+	14.4	64.4	>100^d	75.0	>100^d
11	−	−	−	−	+	—	—	—	—	45.0
12	+	+	+	+	+	5.9	39.2	70.7	15.7	>100^d
13	−	+	+	+	+	—	>100^d	86.4	71.1	7.5
14	−	−	−	+	+	—	—	—	>100^d	>100^d
15	+	+	−	+	+	>100^d	—	—	64.7	10.9
16	−	−	−	+	+	—	—	—	>100^d	76.4
17	+	+	−	+	+	>100^d	—	—	>100^d	79.9
18	+	−	−	+	+	>100^d	—	—	>100^d	25.5
19	−	−	−	+	−	—	—	—	>100^d	—
20	+	−	−	+	−	—	—	—	>100^d	—

a Microtox (M): 45% v/v; Daphnia (D): 100% v/v; Thamnocephalus (T): 100% v/v; Hydra (H): 100% v/v; Selenastrum (S): 91%.

b Microtox (M): 15-min IC$_{50}$; Daphnia (D): 48-h EC$_{50}$; Thamnocephalus (T): 24-h LC$_{50}$; Hydra (H): 96-h EC$_{50}$; Selenastrum (S): 72-h IC$_{50}$.

c Based on sublethal effect endpoint determination.

d Concentration-responses were observed but the 50% effect concentration value was greater than 100% v/v.

Table 7.3. Comparative Bioassay Sensitivity in Detecting and Quantifying Effluent Toxicity.

Sensitivity Ratio	Microtox	Daphnia	Thamnocephalus	Hydra (Sublethality)	Selenastrum
Qualitative[a]	60%	50%	40%	80%	85%
Quantitative[b]	25%	30%	30%	50%	65%
TU[c]	0–16.9	0–24.4	0–7.8	0–6.7	0–37

[a] Number of effluent samples demonstrating positive responses for a particular bioassay *divided by* the total number of effluent samples (n = 20) x 100 (based on Table 7.2 data).

[b] Number of effluent samples for which a 50% endpoint effect concentration was determined with a particular bioassay *divided by* the total number of effluent samples (n = 20) x 100 (based on Table 7.2 data).

[c] Range of toxic units (TU) generated by each bioassay for the 20 effluents.

sured for the 20 effluents span from 0 to 16.9 (Table 7.3). Similar observations of Tables 7.2 and 7.3 data for the other bioassays give insights of their specific and combined responses to the 20 effluents. Overall, on the basis of both qualitative and quantitative responses, bioassay sensitivity decreased in the following order:

Qualitative: *Selenastrum* > *Hydra* > Microtox > *Daphnia* > *Thamnocephalus*
Quantitative: *Selenastrum* > *Hydra* > *Thamnocephalus* = *Daphnia* > Microtox

Although phylogenetic considerations, toxicity levels (e.g., sublethal and lethal responses in this study) and measured endpoints, for example, can offer some explanation for differences seen in bioassay toxicity responses to particular groups of samples, similarities can also be observed.[10,23] In this study, very highly significant correlations ($p < 0.001$) were shown to exist between responses of *Thamnocephalus, Daphnia,* and *Selenastrum,* highly significant correlations ($p < 0.01$) between *Hydra* and Microtox (and *Hydra* and *Thamnocephalus*), and significant correlations ($p < 0.05$) between *Thamnocephalus* and Microtox (Table 7.4). While correlations between *Thamnocephalus* and *Daphnia* are more easily explained by the fact that both organisms share commonality in being freshwater microinvertebrates, similarities in the mode of action of bioactive chemicals present in the effluents in targeting common cellular sites may explain the other correlations. When one considers that a battery of bioassays should ideally represent phylogenetic diversity such that toxicity responses yield nonredundant information,[10] these data suggest that the present battery could be shortened to assess the potential hazards of similar effluents in the future without loss of ecotoxic information. Clearly, the simpler and more cost-effective *Thamnocephalus* assay could be considered a logical alternative to the *Daphnia* assay, especially since both share common phylogenicity. Reducing the test battery further might prove to be unwise, however, as it would decrease phylogenetic diversity and loss of ecotoxic information could conceivably occur when other sets of effluent samples are evaluated.

In a first attempt to identify factors responsible for bioassay toxicity responses, Spearman rank order correlations were undertaken between 50% effect endpoint values (Table 7.2) and effluent physicochemical characteristics (pH, conductivity, DOC and OD_{260} values of Table 7.1). Endpoint values were first transformed into toxic units (as explained earlier) and then correlated to physicochemical data. Significant correlations were observed for three bioassays (Microtox, $p < 0.05$; *Daphnia*, $p < 0.01$; *Hydra*, $p < 0.05$) with conductivity (EC values of Table 7.1), suggesting that their responses might be linked to bioavailable toxic metal concentrations present in the effluents.

In a further attempt to gain insight into toxicity responses more directly linked to the presence of six heavy metals analyzed in the effluents, we explored the toxicity factor (TF) concept explained previously. From toxicity endpoint

Table 7.4. Interbioassay Correlation Coefficient[a] Matrix Based on Toxicity Units Data Obtained with the 20 Effluents.

Bioassay	Microtox	Daphnia	Thamnocephalus	Hydra (Sublethality)	Selenastrum
Microtox	1.000				
Daphnia	0.435	1.000			
Thamnocephalus	0.474*	0.929***	1.000		
Hydra	0.605***	0.421	0.649**	1.000	
Selenastrum	0.257	0.763***	0.785***	0.384	1.000

[a] $p < 0.05$ for *; $p < 0.01$ for **; $p < 0.001$ for *** based on Pearson product-moment correlation coefficient matrix (Zar, 1984).

Table 7.5. Toxicity Responses of Bioassays[a] to Six Heavy Metals Analyzed in the Effluent Samples Studied.

Chemical	Microtox	Daphnia	Thamnocephalus	Hydra (Sublethality)	Selenastrum
Cd^{2+} $(CdCl_2)$	14.5	0.148	0.2	0.854	0.173
Cr^{6+} $(K_2Cr_2O_7)$	104	0.0291	0.018	0.005	0.487
Cu^{2+} $(CuSO_4.5H_2O)$	0.211	0.0328	0.0895	0.441	0.0425
Mn^{7+} $(KMnO_4)$	–	0.13	–	0.428	0.286
Mn_2+ $(MnCl_2)$	75.5	13.9	–	–	1.27
Mn^{2+} $(MnSO_4.H_2O)$			23.3		
$Ni2+$ $(NiCl_2.6H_2O)$	110	1.53	–	0.443	0.187
Ni^{2+} $(NiSO_4.6H_2O)$			2.21		
Zn^{2+} $(ZnSO_4.7H_2O)$	0.872	1.16	0.69	0.446	0.053

[a] Microtox (M): 15-min IC_{50}; Daphnia (D): 48-h EC_{50}; Thamnocephalus (T): 24-h LC_{50}; Hydra (H): 96-h EC_{50}; Selenastrum (S): 72-h IC_{50}.

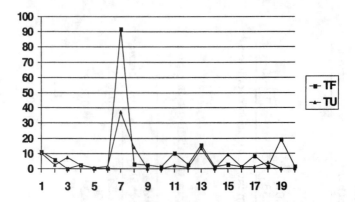

Figure 7.2. TFᵐ values for six metals and TU (toxicity unit) values generated with the algal assay for the 20 effluents. Associations between heavy metal toxicity and algal bioassay responses are suggested for effluents 1, 7, and 13.

values determined for these six metals with each of the bioassays (Table 7.5), we were able to calculate TF_i^m values for each heavy metal and, from these, TF^m values (which reflect the sum of six metal TF_i^m values for a particular bioassay). As an example, Figure 7.2 illustrates TF^m values and toxicity units (TUs) generated with the algal assay for the 20 effluents. Heavy metals appear to have an effect in effluents 1 (TU = 10 and TF^m = 11), 7 (TU = 37.3 and TF^m = 91.8), and 13 (TU = 13.3 and TF^m = 15.5). Upon examination of the breakdown of TF^m into individual TF_i^m values, the latter suggest that Zn (TF^m = 8.5) and Ni (TF^m = 2.5) play a role in effluent 1 toxicity, and that Zn (TF_i^m = 83.8) and Cu (TF_i^m = 7.53) play a role in effluent 7 toxicity (Table 7.6). Similarly, Ni (TF_i^m = 10.4) and Zn (TF_i^m = 3.96) are putative candidates of toxic effects in effluent 13.

Table 7.7 summarizes the results of comparable toxicity factors and toxicity unit evaluations for all effluents. Some of the major points which stand out in this table are highlighted below:

1. Zn and Cu were clearly the main factors responsible for toxicity in effluent 7 (metal processing) and marked toxicity responses were observed in all bioassays.
2. Toxicity responses of effluent 3 (pharmaceuticals) generated with all five bioassays are not associated with metal TF^m values, suggesting that substances other than metals were responsible for toxic effects.
3. In the case of the *Hydra* assay, TF_i^m values for Mn were calculated assuming that this metal was present as MnO_4^{1-} rather than as Mn^{2+} (see Table 7.5). We may therefore have overestimated the overall toxic contribution of Mn in so doing.

Table 7.6. Relative Importance of the Toxicity Factors of Analyzed Metals with Respect to the *Selenastrum* Assay for Effluents 1, 7, and 13.

Effluent	Toxicity Factor (TF_i^m) Contribution of Analyzed Metals					
	Cd	Cr	Cu	Mn	Ni	Zn
1	0	0	0	0	2.5	8.5
7	0	0	7.53	0.49	0	83.8
13	0	0.03	0.05	1.07	10.4	3.96

4. Zn, Cu, Ni, Cr, and Mn (and possibly other nonanalyzed metals) likely exerted toxic stress singularly or in combination in samples where bioassays expressed a toxic response. Cd, in contrast, was certainly not an actor in any toxicity response because of its insignificant presence (maximum analyzed concentration of 1.1 μg/L in effluent 12).

5. Effluents 9 (sewage treatment) and 19 (laundry cleaning facility) displayed relative (low, moderate, or high) toxicity factors (TF^m values), yet showed no toxicity with any of the bioassays. We suspect that the presence of toxicity-modifying factors (e.g., organic substances, specific ions) in these effluents may have reduced metal bioavailability in this case.

To further corroborate the complicity of metals in the bioassay toxicity responses, we conducted ancillary tests with EDTA (ethylenediaminetetraacetic acid), a well-known metal chelating agent.[26] Microtox testing, before and after addition of EDTA, was performed on aqueous solutions of metals analyzed in this study, as well as on selected effluents having previously shown toxicity toward this bioassay (Table 7.8). Four heavy metal solutions (Cd, Cu, Cr, Zn) demonstrated a significant decrease in toxicity following EDTA treatment, as did one of the three effluents tested. Effluent 7 (metal processing) results once again support the hypothesis that Zn and Cu were likely the main causes of toxicity, as purported by TF^m values reported in Table 7.7. In contrast, effluent 10 (textiles) and 12 (chemical production) results exhibited increases in toxicity after EDTA addition, which is contrary to expectations if heavy metals were actually involved in toxic effects. Moreover, since the toxicity of effluent 10 was drastically reduced after pH adjustment (Table 7.8), it appears that its toxicity was entirely pH-related.

In similar EDTA-related experimentation with the *Selenastrum* assay and effluents 1, 7, 8, and 13 (results not shown), a significant toxicity decrease was again observed with effluent 7 (metal processing) and also with effluent 13 (aluminium production), suggesting metal involvement in the toxicity response. While toxicity decreases in the case of effluents 1 (pulp and paper) and 8 (meat processing) were not significant, they nevertheless suggest some metal influence on toxicity expression, possibly linked to Zn (see Table 7.7).

Table 7.7. Contribution[a] of Heavy Metals[b] to Effluent Toxicity.

Effluent	Microtox	*Daphnia*	*Thamno.*	*Hydra*[c]	*Selenastrum*
#1		Ni		Mn>Ni>Zn	Zn>Ni
#2				Zn	Zn
#3					
#4		Cu			Zn>Cu
#5				Mn	
#6		Cr	Cr		
#7	Zn>Cu	Cu>Zn	Zn>Cu	Zn>Mn>Cu	Zn>Cu
#8			Cr	Mn>Cr	Zn>Mn
#9		Cr>Cu	Cr>Cu	Cr>Mn	Mn>Cu
#10			Cr	Mn>Cr	Mn
#11			Cr>Zn	Mn>Cr>Zn	Zn>Mn
#12			Cr	Mn>Cr	Mn
#13		Ni>Cr	Cr>Ni	Ni>Cr>Mn	Ni>Zn>Mn
#14				Mn	Mn
#15			Cr	Mn>Cr	Mn>Zn
#16				Mn	Mn
#17		Cr	Cr>Zn	Cr>Mn>Zn	Zn>Mn
#18				Mn	Mn
#19	Zn		Zn>Cr	Cr>Mn>Zn	Zn>Mn
#20		Cu>Cr	Cr	Cr>Mn	Mn

[a] Rectangular boxes indicate range of toxic responses (in toxicity units = TU) generated by each bioassay: dotted box ($1 > TU > 0$); single frame box ($5 > TU > 1$); double-framed box ($TU > 5$).

[b] Metals appear in boxes when their respective Toxicity Factor (TF) exceeded a value of 0.5: *underlined metal* ($1 > TF > 0.5$); metal in normal letters ($5 > TF > 1$); **metal in bold letters** ($TF > 5$).

[c] TF of Mn was calculated based on EC_{50} of Mn^{7+} ($KMnO_4$).

Table 7.8. Effect of EDTA[a] on Microtox Toxicity for Selected Metals and Effluents.

| Sample | Microtox 15-min IC50s (95% Confidence Intervals) in mg/L for Metals and in % v/v for Effluents | | |
	Before Addition of EDTA	After Addition of EDTA	Resulting Effect on Toxicity
Cd	14.5 (11.1–19.0)	27.7 (24.7–31.2)	decrease
Cu	0.211 (0.15–0.29)	7.11 (6.68–7.61)	decrease
Cr	104.1 (50.8–213)	>100	decrease
Mn	75.5 (62.7–90.8)	55.2 (33.3–91.6)	no significant change
Ni	110 (78.1–155)	61.5 (49.8–75.9)	increase in toxicity
Zn	0.872 (0.63–1.21)	2.65 (1.44–4.86)	decrease
Effluent 7	10.2 (9.02–11.6)	49.7 (33.1–74.7)	decrease
Effluent 10	43.8[b] (43.4–44.3)	21.7 (21.4–21.9)	increase
Effluent 12	5.9 (5.58–6.22)	4.0 (3.71–4.24)	increase

[a] 90.4 mg/L of EDTA (previously determined as the No Observed Effect Concentration for the Microtox testing system) was added to each sample.

[b] An extrapolated 15-min IC_{50} value of 146%#v/v (85.8–249% v/v) was determined after sample pH was neutralized from 9.7 to 6.9.

Time-related responses in certain bioassays offer yet another means of obtaining information on putative causes of toxic effects. Microtox 5-, 15-, and 30-minute IC50s, for example, were shown to decrease significantly with time of exposure for a metal plating effluent, while similar IC50s determined for a pesticide production plant effluent demonstrated no time-related changes.[27] Slow-reacting metals in the Microtox test such as Cd, Cu, Zn, and Ni are known to enhance the sensitivity of the luminescent bacterial reagent as time of exposure augments.[28,29] In this work, effluent 7 (metal processing) yielded (significantly different) Microtox 5-, 15-, and 30-minute IC50s of 16.8, 10.2, and 7.6 %v/v, respectively, which again suggests the presence of metallic toxicity and substantiates the toxic factor and EDTA-related experimental results.

The *Daphnia* immobilization assay, conducted on a sharply odoriferous (ammonia-like odor) effluent 3 (pharmaceuticals) sample, yielded (significantly different) 24- and 48-hours EC50s of 51.8 and 99.2% v/v, respectively, suggestive of the presence of volatile organic toxicant(s). The second pharmaceuticals wastewater, effluent 5, had shown sublethal effects in the *Hydra* assay after 24- ($EC_{50} = 67.4\%$ v/v) and 48-hours ($EC_{50} = 60.9\%$ v/v), but these had disappeared at the 72- and 96-hours times of exposure (EC50s > 100% v/v). Since *Hydra* is not known to possess strong capabilities for xenobiotic metabolization,[30] these time-related evanescent sublethal effects again suggest the presence of some volatile organic toxicity in effluent 5.

Finally, PEEP (Potential Ecotoxic Effects Probe) index values were determined in order to estimate the degree of noxiousness of each of the 20 effluents on a common comparative basis with the help of the PEEP formula.[10] Each PEEP value is simply the $\log_{10}$ of an effluent's toxic loading (= product of the average sum of toxic units generated by bioassays *and* the effluent's flow expressed in m^3/h) and reflects the specific relative toxic potential of each wastewater in relation to the group (n = 20). As an example, effluent 1 displayed a « toxic print » of 38.1 (i.e., toxic units calculated from the « n $(\Sigma$ Ti)/N » part of the formula) and had a flow « Q » of 100,000 m^3/d (Table 7.1) or 4167 m^3/h. The product of « n $(\Sigma$ Ti)/N » and « Q », expressed in m^3/h, yields a toxic loading value of 158,763, the $\log_{10}$ (or PEEP value) of which is 5.2. While bioassays, in this study, were not repeated on each of the 20 Japanese effluents after a biodegradation step as suggested in the original PEEP article,[10] this does not preclude utilization of the PEEP concept as applied here. Calculated PEEP values clearly demonstrate a wide spectrum of toxic activity for the effluents, as they range from 0 to 5.2 and therefore indicate that some have the potential to cause much more harm to receiving water biota than others (Figure 7.3). Effluents 1, 3, 12, and 13 markedly contrast with effluents 9, 14, and 19 in terms of toxic strength. While the latter are characterized by PEEP values of 0 (revealing an absence of measurable responses from all five bioassays) and do not contribute any toxic loading to the receiving environment, the former display PEEP values of 5.2, 5.2, 5.2, and 5.1, respectively, and make up 87.5% (23% + 23% + 23% + 18.5%) of the total toxic loading of the group of 20 effluents. If reduction of wastewater toxic input should become a desired objective for this group of effluents to sustain the quality of receiving aquatic ecosystems, it is clear that effluent amelioration efforts should be first directed toward those industrial plants harboring the highest PEEP values (i.e., highest toxic loading). For such a purpose, the value of this bioassay-directed PEEP scale as a management tool to aid in decision-making in order to prioritize curative actions relating to industrial point source pollution can certainly be appreciated.

At this time, PEEP index values in the second author's Centre Saint-Laurent laboratories have been generated for 77 industrial effluents discharging into the Saint-Lawrence River under the first (1988–1993) and second (1993–1998) Saint-Lawrence River Action Plans (Costan et al., 1993; Bermingham et al., 1994; and more recent unpublished results). These liquid wastes stem from diverse industrial sectors (e.g., pulp and paper, metallurgy plants, petroleum refinery, textile, organic and inorganic chemical production plants) and displayed PEEP values ranging from 0 to 7.5. When PEEP values for these effluents are recalculated

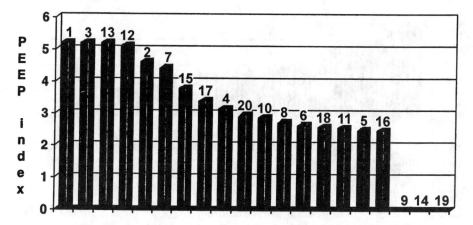

Figure 7.3. PEEP values for the 20 Japanese-based effluents investigated. Numbers associated with PEEP bar values identify effluents listed in Table 7.1.

without the postbiodegradation step bioassay results in order to make them more comparable to the Japanese effluent PEEP values, PEEP values are virtually unchanged and range from 0 to 7.6 (Figure 7.4). Within this group of effluents, six contribute 91.9% of the total toxic loading with respective PEEP values of 7.6 (one inorganic chemical production plant effluent: 27.7% of toxic loading), 7.5, 7.4, and 7.2 (three pulp and paper plant effluents: 23.4%, 19.1%, 12.0% of toxic loading, respectively), 6.9 (one inorganic chemical production plant: 5.4% of toxic loading) and 6.8 (one metallurgical plant effluent: 4.3% of toxic loading).

While the group of 77 effluents studied under the St-Lawrence River Action Plans (PEEP value range from 0 to 7.6) and the group of 20 Japanese effluents (PEEP value range from 0 to 5.2) are not strictly comparable because toxicity data were not generated with exactly the same bioassays and endpoints, it is nevertheless of interest to note that the two groups vary by more than 2 orders of magnitude. Canadian-based effluents displaying PEEP values exceeding those of the highest values calculated with the Japanese effluents (i.e., PEEP value > 5.2) came from twelve pulp and paper plants, two inorganic chemical production plants, one textiles plant, one metallurgical plant and one sewage treatment works plant (mixed domestic and industrial wastes). At the time of their ecotoxicological assessment (1989–1992), these Canadian plants did not have optimal waste treatment systems in operation, which may partially explain their very high PEEP values. In contrast, the group of Japanese-based plants we investigated in 1995 invariably had in place secondary treatment (or equivalent) for all 20 effluents (Table 7.1). The importance of secondary treatment as a means of effectively reducing effluent toxicity, particularly with respect to the pulp and paper industry, is now well recognized.[31] Barring the 17 Canadian-based effluents character-

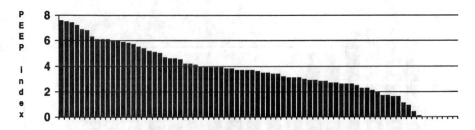

Figure 7.4. PEEP values for 77 Canadian-based effluents investigated under two Saint-Lawrence River Action Plans (see text).

ized by PEEP values above 5.2, the remaining 60 (or 78%) displayed a PEEP range commensurate with that of the Japanese-based effluents (Figure 7.4). In-depth multivariate statistical analyses are planned with these two groups of efflu-ents in the near future to better understand cause and effect relationships between their chemical (contaminants) concentrations and individual bioassay responses.

In conclusion, this study will have shown the usefulness of combining a suite of small-scale bioassays and chemical analyses to detect and quantify the degree of toxicity associated with a select group of wastewaters discharged by plants repre-sentative of diverse industrial sectors. Clearly, biology (effects monitoring with bioassays) and chemistry (identification of putative bioavailable chemicals linked to bioassay responses) offer a practical and relevant approach to investigate the toxic potential of point source pollution. Furthermore, when investigating a group of liquid wastes in an attempt to understand which ones may pose the greatest threat to receiving water biota in terms of toxic loading, integrating bioassay re-sponses and effluent flow via the PEEP scale concept reported herein is an effec-tive means of identifying problematic discharges. Although toxicity problems linked to industrial wastes are being increasingly and effectively addressed in many parts of the world, the battle to curb and eradicate harmful contaminants from air, soil, and water compartments of our biosphere is far from being won. It is expected that cost-effective strategies employing bioassay assemblages and chemical analy-sis will continue to be used well into the twenty-first century to ensure that indus-trial processes do not impose undue hazard and risk on our environment.

ACKNOWLEDGMENTS

This Canada/Japan cooperative venture in the field of applied ecotoxicology was undertaken while the second author was on special assignment at the College of Technology, Toyama Prefectural University, Toyama, Japan, during the fall of 1995. Grants obtained from the Japan Society for the Promotion of Science (Foreign Fellowship Division, Tokyo) *and* the Japan Science and Technology

Fund (a component of the Pacific 2000 strategy jointly managed by External Affairs and International Trade Canada, Industry, Science and Technology Canada and the National Sciences and Engineering Research Council of Canada) provided the essential financial support required for ensuring this bilateral initiative. The authors are also indebted to their respective managements for supporting this research project. Appreciation is extended as well to Dr. N. Bermingham, Centre Saint-Laurent, for providing data and useful commentary relating to industrial effluents investigated under the Saint-Lawrence River Action Plans.

REFERENCES

1. Houk, V.S. The genotoxicity of industrial wastes and effluents: A review. *Mutation Res.*, 277, pp. 91–138, 1992.
2. Garric, J., E. Vindimian, and J.F. Férard. Ecotoxicology and wastewater: Some practical applications. *Sci. Total Environ.*, Supplement (Part 2), pp. 1085–1103, 1993.
3. Grothe, D.R., K.L. Dickson, and D.K. Reed-Judkins, Eds. *Whole Effluent Toxicity Testing: An Evaluation of Methods and Prediction of Receiving System Impacts*. SETAC Pellston Workshop on Whole Effluent Toxicity, 1995 Sept. 16–25, Pellston, MI. SETAC Press, Pensacola, FL, 1996, p. 340.
4. USEPA (U.S. Environmental Protection Agency). Technical Support Document for Water Quality-Based Toxics Control. Office of Water, Washington, DC, EN-336, 1985.
5. OECD (Organisation for Economic Cooperation and Development). The Use of Biological Tests for Water Pollution Assessment and Control. OECD Environment Monographs, No. 11, October 1987, p. 70.
6. Blaise, C., G. Sergy, P. Wells, N. Bermigham, and R. van Coillie. Biological testing—Development, application and trends in Canadian environmental protection laboratories. *Toxicity Assessment*, 3, pp. 385–406, 1988.
7. Stulfauth, T. Ecotoxicological Monitoring of Industrial Effluents, in *Environmental Toxicology Assessment*. Richardson, M., Ed., Taylor & Francis, London, UK, 1995.
8. Dutka, B.J. and K.K. Kwan. Battery of screening tests approach to sediment extracts. *Toxicity Assessment*, 3, pp. 303–314, 1988.
9. Blaise, C. Microbiotests in aquatic ecotoxicology: characteristics, utility, and prospects. *Toxicity Assessment*, 6, pp. 145–155, 1991.
10. Costan, G., N. Bermingham, C. Blaise, and J.F. Férard. Potential ecotoxic effects probe (PEEP): A novel index to assess and compare the toxic potential of industrial effluents. *Environ. Toxicol. Water Qual.*, 8, pp. 115–140. 1993.
11. Wells, P., K. Lee, and C. Blaise, Eds. *Microscale Testing in Aquatic Toxicology— Advances, Techniques and Practice*. CRC Press, Boca Raton, FL, 1998, p. 679.

12. Bermingham, N., D. Boudreau, and G. Costan. Synthèse de l'application du barème d'effets écotoxiques potentiels (BEEP) et recommandations d'orientation pour son développement. Centre Saint-Laurent, Conservation de l'Environnement, Région du Québec, Environnement Canada, March 1994, p. 30.

13. JSWA (Japan Sewage Works Association). Wastewater Examination Methods, 1984.

14. Bulich, A.A., M.W. Greene, and D.L. Isenberg. Reliability of the bacterial luminescence assay for determination of the toxicity of pure compounds and complex effluents, in *Aquatic Toxicity and Hazard Assessment*, Branson, D.R. and K.L. Dickson, Eds., Fourth Conference of the American Society for Testing and Materials, STP737, Philadelphia, PA, 1981.

15. Environment Canada. Biological Test Method: Toxicity Test Using Luminescent Bacteria (*Photobacterium phosphoreum*), Environmental Protection Series, Environment Canada, Report EPS 1/RM/24, November 1992, p. 61.

16. Blaise, C., R. Legault, N. Bermingham, R. van Coillie, and P. Vasseur. A simple microplate algal assay technique for aquatic toxicity assessment. *Toxicity Assessment*, 1, pp. 261–281, 1986.

17. Environment Canada. Biological Test Method: Growth Inhibition Test Using the Freshwater Alga *Selenastrum capricornutum*, Report EPS 1/RM/25, Conservation and Protection, Environment Canada, Ottawa, Canada, 1992, p. 42.

18. Blaise, C., J.F. Férard, and P. Vasseur. Microplate Toxicity Tests with Microalgae: A Review, in *Microscale Testing in Aquatic Toxicology: Advances, Techniques and Practice*, Wells, P., K. Lee, and C. Blaise, Eds., CRC Press, Boca Raton, FL, 1998.

19. Cabridenc, R. and P. Lundhal. Intérêt et limites du test daphnie pour l'étude des nuisances des polluants. *Techniques et Sciences Municipales de l'Eau*, 6, pp. 340–345, 1974.

20. Environment Canada. Biological Test Method: Reference Method for Determining Acute Lethality of Effluents to *Daphnia magna*, Environmental Protection Series, Environment Canada, Report EPS 1/RM/14, July 1990, p. 18.

21. Creative Selling. Thamnotoxkit™ F. Crustacean Toxicity Screening Test for Freshwater. Standard Operational Procedure. V241092, 1992, p. 23.

22. Blaise, C. and T. Kusui. Acute toxicity assessment of industrial effluents with a microplate-based *Hydra attenuata* assay. *Environ. Toxicol. Water Qual.*, 12, pp. 53–60, 1997.

23. Côté, C., C. Blaise, J.-R. Michaud, L. Ménard, S. Trottier, F. Gagné, P. Riebel, and R. Lifschitz. Comparisons between micro-scale and whole sediment assays for freshwater sediment toxicity assessment. *Environ. Toxicol. Water Qual.*, 13, pp. 93–110, 1998.

24. Sprague, J.B. and B.A. Ramsay. Lethal levels of mixed copper-zinc solutions for juvenile salmon. *J. Fish. Res. Bd. Can.*, 22, pp. 425–432, 1965.

25. Zar, J.H. *Biostatistical Analysis.* Prentice-Hall, Englewoods Cliffs, NJ, 1984, p. 718.

26. Miller, W.E., J. Greene, and T. Shiroyama. The *Selenastrum capricornutum* Printz Algal Assay Bottle Test: Experimental Design, Application, and Data Interpretation Protocol. U.S. Environmental Protection Agency Report No. EPA-600/9-78-018, Corvallis, OR, 1978, p. 126.

27. Calleja, A., J.M. Baldasano, and A. Mulet. Toxicity analysis of leachates from hazardous wastes *via* Microtox and *Daphnia magna. Toxicity Assessment,* 1, pp. 73–83, 1986.

28. Codina, J.C., A. Pérez-Garcia, P. Romero, and A.D. Vicente. A comparison of microbial bioassays for the detection of metal toxicity. *Arch. Environ. Contam. Toxicol.,* 25, pp. 250–254. 1993.

29. Blaise, C., R. Forghani, R. Legault, J. Guzzo, and M. Dubow. A bacterial toxicity assay performed with microplates, microluminometry and Microtox® reagent. *Biotechniques,* 16, pp. 932–937, 1994.

30. Fu, L.J., R.E. Staples, and R.G. Stahl. Assessing acute toxicities of pre- and post-treatment industrial wastewaters with *Hydra attenuata*: A comparative study of acute toxicity with the fathead minnow, *Pimephales promelas. Environ. Toxicol. Chem.,* 13, pp. 563–569, 1994.

31. Blaise, C., R. van Coillie, N. Bermingham, and G. Coulombe. Comparaison des réponses toxiques de trois indicateurs biologiques (bactéries, algues, poissons) exposés à des effluents de fabriques de pâtes et papiers. *Revue Internationale des Sciences de l'Eau,* 3, pp. 9–17, 1987.

Chapter 8

Bioremediation of Freshwater Sediments Contaminated with Polycyclic Aromatic Hydrocarbons

A.G. Seech and J.T. Trevors

INTRODUCTION

Throughout the world, diverse industrial and agricultural activities have resulted in deposition of toxic organic compounds in freshwater ecosystems. The extent of sediment contamination in many industrialized regions is so great that following their removal by dredging, sediments from many rivers and harbors must be handled, stored, and disposed of, as hazardous wastes. Extensive contamination of sediments is a significant concern because of its detrimental impact on the aquatic ecosystem and, ultimately, society as a whole.

As regulatory agencies have imposed increasingly strict controls on the release of contaminants to ecosystems by regulating the quality of water which can be discharged from municipal and industrial sources, the focus has shifted toward the issue of contamination already present in the ecosystems' sediments. The key issues are if contaminated ecosystems will recover naturally[1] and the appropriate means by which anthropogenic intervention may be used to accelerate ecosystem remediation.[2] The observation that many organic contaminants have long half-lives in sediments has caused governments in many countries to initiate programs to characterize, remove, and remediate or dispose of contaminated sediments present in many aquatic ecosystems. Polycyclic aromatic hydrocarbons (PAHs) are among the most widely distributed contaminants in freshwater sediments, and a number of PAH compounds are carcinogenic, mutagenic, or teratogenic, and hence, can adversely influence ecosystem health. This chapter focuses on bioremediation of freshwater sediments contaminated with PAHs.

BIODEGRADATION OF PAHs IN SEDIMENT

One of the major fates of organic contaminants in both freshwater and marine sediments is biodegradation. The rate at which biodegradation proceeds is controlled by interacting physical and chemical properties of the contaminants and the sediments in which they reside. PAHs are one of the most common contaminants in freshwater sediments. For example, elevated concentrations of PAHs have been found in sediments in many industrialized areas.[3-5] In situ biodegradation of aromatic compounds including PAHs in anaerobic sediments is known to occur;[6,7] however, degradation proceeds slowly because of low temperatures and limited oxygen availability. In fact, extrapolation of natural PAH degradation rates observed under anoxic conditions indicates that hundreds to thousands of years would pass before natural microbial processes would result in significant reductions in PAH concentrations. Thus, dredging of sediments has been initiated in many areas as a first step toward remediation of the aquatic ecosystems.[2,8] Once the sediments have been removed by dredging it becomes feasible to apply bioremediation as a means of removing organic contaminants. This chapter provides a brief review of the factors which influence biodegradation of PAHs in sediment, case studies on bench-scale and pilot-scale bioremediation of freshwater sediments contaminated with PAHs, and an examination of the extent to which bioremediation can reduce sediment toxicity.

Following compounds composed of glucose residues, those based on the benzene ring structure are the most common in the biosphere.[9] Furthermore, many common PAHs are known to occur naturally in the environment as a result of forest fires and volcanic eruptions.[10] Thus, the observation that many microorganisms have evolved enzyme systems for PAH catabolism is not surprising. In fact, complete aerobic biodegradation of many PAHs to carbon dioxide has been extensively documented in terrestrial environments.[11-15] In sediments, however, biodegradation of PAHs is less extensive because of physical and chemical factors that restrict aerobic microbial activity.

Influence of Oxygen Availability on Biodegradation of PAHs

Central among factors regulating the rate of PAH degradation in sediments is the availability of oxygen. Empirical evidence in support of this observation is found in the recalcitrance of low molecular weight PAHs, which are rapidly biodegraded in aerobic environments, in oxygen stressed ecosystems.[11] The results of studies conducted by Gardner et al.[16] which revealed that biodegradation of PAHs was substantially more rapid near the sediment/water interface than deeper in the sediment layer, in spite of the presence of high bacterial numbers and adequate nutrients, lends additional support. Microcosm studies conducted in our labora-

Table 8.1. Influence of Depth Within Sediment[a] on Mineralization of ^{14}C-Phenanthrene During 7 Days Incubation, at 20°C.

Incubation Conditions	Mineralization (%)			
	0–24 h	24–48 h	48–120 h	120–168 h
Air-dry sediment at a moisture content of 7% (w/w), supplemented with $HgCl_2$ at 2% (w/w)	0	0	0	0
Saturated sediment,[b] ^{14}C-phenanthrene deposited at the sediment/water interface	9.7	11.9	14.1	15.4
Saturated sediment, ^{14}C-phenanthrene deposited 3 cm below the sediment/water interface	5.9	7.3	8.9	9.9

[a] PAH-impacted sediment collected from Thunder Bay Harbour on the north shore of Lake Superior, Ontario, Canada (initial PAH concentration: 1,700 mg/kg; pH: 7.1; texture: silty clay; total organic carbon: 4.2%; total nitrogen: 0.08%).

[b] Sediment at 110% of water holding capacity. A thin layer of free water, approximately 0.2 cm thick, was present on the sediment's surface.

tory, in which mineralization of phenanthrene was monitored at the sediment surface and deeper within the same sediment, provided similar results. Specifically, 15.4% of the ^{14}C-phenanthrene deposited at the sediment/water interface was mineralized to $^{14}CO_2$ during seven days incubation at 20°C; however, when ^{14}C-phenanthrene was deposited at a point approximately 3 cm below the sediment/water interface only 9.9% was mineralized in seven days (Table 8.1).

These findings are not surprising, since sediments often contain high numbers of heterotrophic microorganisms[17] which create a high biological demand for oxygen. In such environments microbial oxygen consumption at points below the sediment:water interface will usually exceed the diffusive supply and, hence, anoxia will increasingly dominate as depth of the sediment layer increases. In fact, redox potentials as low as −450 mV have been observed in sediments rich in organic matter.[13] It should also be recognized that under anoxic conditions molecules other than oxygen, including nitrate, sulfate, and carbon dioxide, must serve as electron acceptors and, hence, the energy yielded from catabolism of organic substrates will be substantially lower. For this reason, microbial growth and the rate of PAH biodegradation will be slower in oxygen limited environments.

PAHs in sediments are biodegraded most extensively by aerobic, or facultatively anaerobic, heterotrophic bacteria. The essential first step in the process, incorporation of molecular oxygen as a substituent on the benzene ring structure,[18] is mediated by dioxygenases.[14] It is apparent that a supply of molecular oxygen is prerequisite to most PAH biodegradation, and when the supply of oxygen is limited, the process is inhibited. Thus, rate of in situ biodegradation of PAHs in sediments is commonly quite low as a result of limited availability of oxygen. The low availability of oxygen in freshwater sediments is the main reason why contaminated sediments are commonly dredged prior to initiation of remediation efforts.

Influence of Temperature on Biodegradation of PAHs

Temperature is one of the most important factors regulating microbial activity, including biodegradation of PAHs, in freshwater sediments. As temperature increases above, or falls below, the optimum for a given microorganism, the growth rate is reduced and some enzyme functions are more easily inhibited.[19] In general, the rate of enzymatic activity will double with each 10°C increase in temperature, up to the point where enzyme proteins begin to denature.[20] Thus, low temperatures will reduce the rate of PAH biodegradation even in the presence of adequate oxygen and other required nutrients. The low temperatures that predominate in many temperate region freshwater sediments are also a major reason why contaminated sediments are commonly dredged prior to initiation of remediation efforts.

BIOREMEDIATION OF CONTAMINATED SEDIMENTS

Over the past two decades the public, the scientific community, and regulatory agencies have become increasingly concerned about the impact of contaminated sediments on benthic organisms. This concern has led to extensive study and characterization of sediments in harbors near industrialized areas. As an example, such study has identified areas of 42 harbors on the Great Lakes as areas of concern with respect to contamination.[21] Characterization of contaminated sediments sometimes identifies them as dangerous to the ecosystem on the basis of toxicity. In such cases the contaminated sediments may be dredged to reduce the risk of continued toxicological effects, and the potential for dispersal of contaminants. In addition, contaminated sediments in many areas must be dredged to allow for passage of ships through harbors and rivers. Once contaminated sediments have been dredged they must be disposed of, or remediated, and hence, technologies effective for treatment of dredged sediment are currently being sought.[2] One of the most promising technologies presently under evaluation, for remediation of sediments contaminated with PAHs, is bioremediation, which utilizes the activity of microorganisms to degrade contaminants in sediment. The technology is seen as advantageous as compared to disposal options such as landfilling or encapsulation because it can reduce or eliminate long-term liability through complete decomposition of the target compounds to acceptable residual levels. In addition, bioremediation is often the most attractive alternative from the perspective of cost. The latter consideration can be of considerable importance, since hundreds of thousands of tonnes of contaminated dredged sediment are in need of disposal or remediation.[22]

Since every sediment has a unique set of physical, chemical, and biological characteristics, and these can strongly influence the feasibility and efficiency of bioremediation treatments, it is important that bioremediation protocols be optimized at laboratory scale prior to initiation of large-scale sediment bioremediation. Such laboratory studies are commonly known as treatability investigations, and are designed to (a) characterize the sediment with respect to the physical/chemical properties known to affect biodegradation of the target compounds, (b) determine the biodegradation rates for the target compounds, (c) determine the attainable residual concentrations of the target compounds, (d) determine the major environmental fates of the target compounds, and (e) estimate the influence of treatment on sediment toxicity. The following section provides an overview of a treatability investigation conducted on sediment dredged from Thunder Bay Harbour, near Thunder Bay, Ontario, Canada.

BENCH-SCALE STUDIES

Materials and Methods

Sediment Collection and Preparation

Sediment was collected from Thunder Bay Harbour, near Thunder Bay, on Lake Superior, by Environment Canada. The sediment sample, which weighed approximately 27 kg, was placed in a 22.7 L polyethylene pail with an airtight lid, and transported to the Grace laboratory in Mississauga. Upon receipt in Mississauga the sample was stored at 4°C in the dark. Preparation of the sediment sample for the treatability investigation was initiated nine days later by decanting approximately 0.5 kg of supernatant water. The sediment was spread in a prewashed stainless steel pan under a fume hood to air-dry for a period of 48 hours. A 4.4 kg sample of sediment was removed from the pan prior to drying, placed in a 4.5 L pail, and stored in the cold room for use in subsequent toxicity analyses. The air-dried sediment was screened to less than 2.0 mm. Sediment aggregates larger than 2.0 mm were manually disrupted sufficiently to allow all fragments to pass through the sieve. Following sieving, the sediment was thoroughly homogenized in a planetary motion blender, and triplicate samples were submitted for physical and chemical analyses, including pH, available P, total nitrogen, ammonium, nitrate, total organic C, texture, dichloromethane-extractable oil and grease, chlorinated phenols, PAHs, dioxins, furans, and heavy metals. The prepared sediment was placed in a clean 22.7 L polyethylene pail and returned to 4°C storage until needed for preparation of the treatability microcosms.

Radioisotope Fate Studies Using ^{14}C-benzo(a)pyrene

The treatability investigation was conducted in microcosms consisting of 1.0 L wide-mouth, screw-top, glass jars (Consumers Glass, Mississauga, Ontario; Canada; Figure 8.1). Microcosms received either 300 g (controls) or 550 g (treatments) of air-dry sediment. Treatments were applied to the air-dry sediment as described in Table 8.2. Sediment in each microcosm was supplemented with ^{14}C-benzo(a)pyrene at 1.0×10^6 dpm/microcosm, to allow determination of the influence of treatments on the fate of the compound. The radioisotope was added to soil in microcosms as an ethanol solution via a syringe. Complete biodegradation of benzo(a)pyrene to carbon dioxide (aerobic mineralization) was estimated by monitoring the conversion of ^{14}C-benzo(a)pyrene to ^{14}C-CO_2. Total volume of the radioisotope solution added to soil in each microcosm was between 200 μL and 400 μL. Radioactive CO_2 evolved from soil in microcosms was trapped in concentrated NaOH (20 mL, 2.0 M) placed in a separate vessel within the microcosms, and quantified using a Beckman model LS 6500 (Beckman Instruments, Fullerton CA) liquid scintillation counter. Samples were prepared for counting by mixing a 1.0 mL aliquot of NaOH from each biometer with 1.0 mL of acetic acid and 10.0 mL of scintillation cocktail (Scintiverse, Fisher Scientific, Toronto, Canada).

Mineralization data are presented as percent recovery of the added ^{14}C as $^{14}CO_2$. All treatments were performed in triplicate, unless otherwise indicated, and least significant differences were determined at the 95% confidence level using a standard analysis of variance (ANOVA) procedure. The jars were sealed with Teflon-lined polyethylene lids, and incubated in the dark at 28±1°C. During the period of maintenance of each set of microcosms ^{14}C-CO_2 was quantified on 12 occasions by counting the ^{14}C activity in the resident NaOH, then refilling the traps with fresh NaOH. The ^{14}C-CO_2 is a direct quantification of the ^{14}C radioisotope spiked to the treated soil that has been mineralized. Carbon traps maintained within the microcosms were counted once at the conclusion of the study to quantify the cumulative release of ^{14}C from the added benzo(a)pyrene as volatile organic compounds. In addition, at intervals through the investigation, samples were extracted from selected treatments and air-dried control microcosms and submitted for chemical analysis.

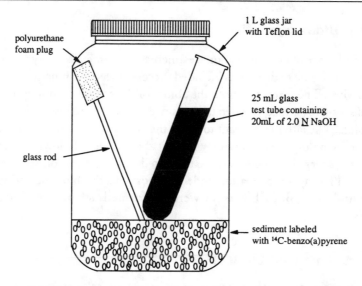

Figure 8.1. Schematic diagram of the microcosm used in radioisotope fate studies.

Table 8.2. Control and Treatments Applied to Sediment in Microcosms Used for the Treatability Investigation.

Control/ Treatment	Amendment 1[a,b]	Amendment 2[a,b]	Amendment 3[a,b]	Moisture Content
Control	none	none	none	air-dry
Treatment 1	D6380, 5% (w/w)	none	D6380, 3% (w/w)	60% WHC
Treatment 2	D6380, 5% (w/w)	triple superphosphate, 0.2% (w/w)	triple superphosphate, 0.07% (w/w)	60% WHC
Treatment 3	D6380, 5% (w/w)	monoammonium phosphate, 0.2% (w/w)	none	60% WHC
Treatment 4	D6386AJAW, 5% (w/w)	calcium peroxide, 1% (w/w)	none	60% WHC
Treatment 5	D6386, 3% (w/w)	none	triple superphosphate, 0.07% (w/w)	60% WHC

[a] D6380 and D6386 are identifiers for DARAMEND® bioremediation products, which are solid-phase organic amendments manufactured by comminution and processing of natural botanical materials (Grace Bioremediation Technologies, Mississauga, Ontario). Triple superphosphate and monoammonium phosphate are nutrient sources. WHC is the amended sediment's water-holding capacity. Amendment additions are on a weight/weight basis 2.

[b] Amendments 1 and 2 were added at initiation of the treatability investigation; Amendment 3 was added on day 79.

Toxicity Bioassays

The influence of bioremediation treatments on sediment toxicity was assessed using bioassays. The toxicity tests included success of seed germination and earthworm survival. The former assessed the ability of the sediment to support germination of sensitive (lettuce), moderately sensitive (tomato), and insensitive (timothy grass) plants. The latter was meant to assess the ability of the sediment to support survival of living organisms that may inhabit the sediment following treatment. The bioassays were conducted on the untreated sediment and on sediment collected after 113 days of incubation under the most effective bioremediation treatment, number 1. Both bioassays were performed according to standard protocols.[22,23]

RESULTS AND DISCUSSION

Physical and Chemical Characteristics of the Sediment

The results of physical and chemical analyses conducted on the sediment are presented in Table 8.3. The sediment was texturally classified as a silt loam, with a pH of 6.6, and a total organic carbon content of 2.9%. The nutrient profile of the sediment indicated that either available phosphorus or nitrogen could limit biological activity as the total organic carbon:total Kjeldahl nitrogen:available phosphorus ratio was approximately 1,300:60:1, while a more satisfactory ratio would approximate 100:10:1. Due to this potential nutrient limitation, several of the treatments employed included supplemental inorganic phosphorus.

The total PAH concentration of the sediment was 572 mg/kg. The concentration of PAH compounds with 4–6 fused benzene rings, a subset which includes all the carcinogenic and potentially carcinogenic species, was 296 mg/kg. Concentrations of a number of the individual PAH compounds exceeded the standards for industrial soil, as set by the federal environmental agency. The sediment had an initial oil and grease content of 35,000 mg/kg.

Other organic contaminants, including chlorinated phenols, dioxins, and furans, were present at relatively low concentrations. For example, the pentachlorophenol concentration was 0.1 mg/kg and the dioxin/furan TEQ was 251 ng/kg. These concentrations are below the federal and provincial criteria for industrial soil.

Mineralization ^{14}C-benzo(a)pyrene

Results of studies on mineralization of ^{14}C added to the sediment as benzo(a)pyrene are presented in Figure 8.2. The composition of treatments ap-

Table 8.3. Initial Physical and Chemical Characteristics of the Prepared Thunder Bay Sediment.

Parameter	Result	CoV (%)	Units
Sand	15.0	2	% wt/wt
Silt	63.3	2	% wt/wt
Clay	21.7	4	% wt/wt
Water-holding capacity	64.8	2	g/100 g
Available phosphorus	22	8	mg/Kg
Total nitrogen	0.129	12	%
Total organic carbon	2.94	1	%
pH	6.6	1	—
Arsenic	>3.3	3	mg/Kg
Chromium	36.9	2	mg/Kg
Copper	42.8	11	mg/Kg
Lead	13.5	2	mg/Kg
Manganese	313	1	mg/Kg
Nickel	27.4	1	mg/Kg
Zinc	97.7	6	mg/Kg
Oil and grease	3.51	3	mg/g
Total PAHs	572	4	mg/Kg
HMW PAHs	296	6	mg/Kg
PCP	0.107	13	mg/Kg
Dioxins and furans (TEQ)	251		ng/Kg

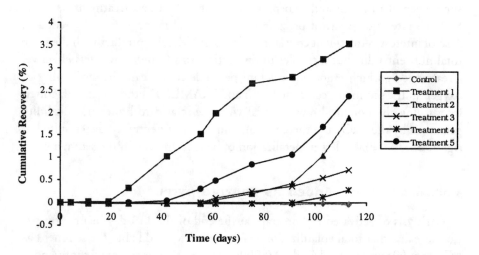

Figure 8.2. Influence of bioremediation treatments on cumulative recovery of ^{14}C as CO_2, adjusted for background radioactivity.

plied to the sediment were provided in Table 8.2. The data on mineralization of ^{14}C-benzo(a)pyrene have been corrected for background radioactivity. In the control sediment, mineralization of ^{14}C-benzo(a)pyrene to CO_2 was not observed over the 113-day incubation period. During the same period, substantial conversion of ^{14}C-benzo(a)pyrene to CO_2 was observed in sediment subjected to bioremediation treatments; however, a lag period prior to the onset of ^{14}C-CO_2 release was observed.

The length of the lag period was influenced by the treatment applied, and ranged from 19 to 91 days. The shortest lag period, at 19 days, was observed in sediment subject to treatment 1, which was composed of an organic amendment, DARAMEND product D6380 without supplemental nutrients (Figure 8.2). In treatments 2 and 3, which included the same organic amendment combined with phosphorus or nitrogen and phosphorus, the length of the lag period was increased (Figure 8.2). Other treatments, including number 4 which utilized an organic amendment in combination with a solid-phase oxygen source (CaO_2), and number 5 which used an organic amendment prepared to have a smaller mean particle size, also resulted in extended lag periods prior to the onset of ^{14}C-benzo(a)pyrene mineralization.

Total mineralization of benzo(a)pyrene and the rate at which it proceeded were sharply influenced by treatments applied to the sediment. The most extensive mineralization, totaling approximately 3.5% over the 113-day incubation, was observed in sediment treated with DARAMEND organic amendment product D6380 (Table 8.2). When sediment was subjected to treatments that combined the same organic amendment with supplemental phosphorus or supplemental nitrogen and phosphorous (treatment 2 and treatment 3, Table 8.2) the rate and extent of benzo(a)pyrene mineralization were both reduced. Also of interest, a treatment consisting of a DARAMEND product with the same total nutrient value but a smaller mean particle size (D6386) was less effective. Treatment 5, which employed calcium peroxide to increase the availability of oxygen in the sediment, combined with a DARAMEND product manufactured by washing the botanical raw material with acetic acid to buffer the alkalinity generated by the CaO_2 and thereby maintain pH near neutrality, also resulted in significantly (p=0.05) less mineralization of benzo(a)pyrene in the sediment.

Volatilization ^{14}C Added as Benzo(a)pyrene

Analysis of activated carbon traps at the end of the 113-day incubation period revealed that total volatilization of ^{14}C, which was added to the sediment as ^{14}C-benzo(a)pyrene, was less than 0.009% and was not significantly influenced by the treatments.

Biodegradation of Native PAHs

Subsamples of sediment from the air-dried control and sediment subjected to bioremediation treatments were submitted for PAH analyses following 14, 27, 42, 57, 91, and 113 days of incubation.

Total PAHs

Total PAHs concentrations decreased for the control and each treatment throughout the 113-maintenance period (Figure 8.3). The greatest reductions in both the control and the treatments were observed during the first 42 days of incubation.

In the control, the total PAH concentration decreased from 572 mg/Kg to 414 mg/Kg in the first 42 days, and further to 390 mg/Kg by day 113. This represented a decrease of 28% in 42 days and 32% after 113 days of incubation. The observed reductions in PAH concentration was probably due to nonbiological effects, including volatilization and reduced extractability of PAHs in the air-dry soil. Since the rate of volatilization is concentration-dependent, and ^{14}C-benzo(a)-pyrene mineralization data indicated that biodegradation resulted in substantial reductions in PAH concentrations in the sediment subjected to bioremediation treatment, it may be assumed that the contribution of volatilization to observed decreases in PAH concentration was greater in the control than for the treatments. Studies conducted in our laboratory (unpublished data) provide strong evidence that the extractability of high molecular weight PAHs falls substantially as the intensity and duration of drying increases.

The most effective treatment, number one, the total PAH concentration was reduced from 572 mg/Kg to 164 mg/Kg during the first 42 days of incubation, representing a decrease of 72%. During the period between days 42 and 113 a further reduction to 73 mg/Kg was observed, resulting in a total reduction of 87% by the end of the investigation. Examination of the data (Figure 8.3) suggests that additional reductions in PAH concentrations could be expected in sediment subjected to bioremediation treatment, albeit at a slower rate (i.e., no treatment plateau had been encountered by day 113 of the investigation). This assumption is supported by data on mineralization of added ^{14}C-benzo(a)pyrene, which indicated that biodegradation of compound was continuing at a substantial rate (Figure 8.2).

Other treatments, which utilized the same organic amendment in combination with triple superphosphate (number 2) or monoammonium phosphate (number 3), resulted in less extensive reductions in PAH concentrations. The second most effective treatment, number two, utilized the same organic amendment supplemented with triple superphosphate (Figure 8.3). Treatment number 5, which

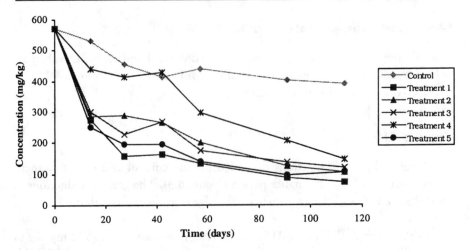

Figure 8.3. Influence of bioremediation treatments on total PAH concentrations.

utilized an organic amendment with the same total nutrient content but with a
smaller mean particle size, applied at a lower dosage (i.e., 3% vs 5% by weight)
with addition of supplemental phosphorus on day 78 of incubation, resulted in
similar reductions in the total PAH concentration. Treatment number 4, which
combined a chemical source of oxygen, CaO_2, with an organic amendment designed to neutralize alkalinity generated by dissolution of the CaO_2, was the least
effective, from the perspective of reducing PAH concentration of the sediment.

High Molecular Weight and Carcinogenic PAHs

High molecular weight (HMW) PAHs may be defined as those containing
4–6 fused benzene rings, specifically fluoranthene, pyrene, benzo(a)anthracene,
chrysene, benzo(g,h,i)perylene, benzo(b)fluoranthene, benzo(k)fluoranthene,
indeno(1,2,3-c,d)pyrene, dibenz(a,h)anthracene, and benzo(a)pyrene. These PAHs
are, in general, less volatile, less water-soluble, and therefore more resistant to
biodegradation that PAHs of lower molecular weight. A number of the HMW
PAHs, including benzo(a)anthracene, benzo(a)pyrene, chrysene, benzo(b)fluoran-
thene, dibenz(a,h)anthracene, indeno(1,2,3-c,d)pyrene, and benzo(g,h,i)perylene
are considered to be carcinogenic.[14] Studies designed to assess the feasibility of
bioremediation as a method of decontaminating PAH-impacted sediments must
focus on the removal of HMW and carcinogenic PAHs because, in general, the
remediation criteria for these compounds are more stringent than those for other
PAHs.

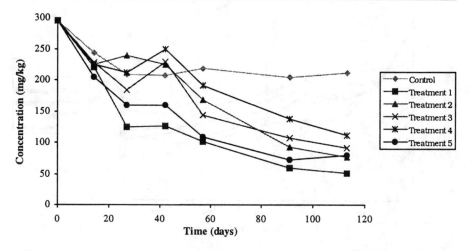

Figure 8.4. Influence of bioremediation treatments on high molecular weight PAH concentrations.

High Molecular Weight PAHs

The concentration of HMW PAHs in the control sediment decreased by 29%, from 296 mg/kg to 208 mg/kg, during the first 27 days of incubation (Figure 8.4). During the remainder of the investigation, from day 27 to day 113, the extractable concentration of HMW PAHs in the control did not change substantially.

Concentrations of HMW PAHs were reduced more substantially in sediment subjected to bioremediation treatments, as compared to the control. In sediment subjected to the most effective treatment, number one, the HMW PAH concentration decreased by 58%, from 296 mg/kg to 125 mg/kg, in the first 27 days of incubation. During the remainder of the investigation, from day 27 to day 113, the extractable concentration of HMW PAHs was further reduced to 51 mg/kg (Figure 8.4), resulting in a cumulative reduction of 83% by the end of the incubation.

Carcinogenic PAHs

In the control sediment, a reduction of 46% in the concentration of carcinogenic PAHs, from 134 mg/kg to 67 mg/kg, was observed in the first 27 days of incubation. During the remainder of the investigation, from day 27 to day 113, no further reduction in the concentration of carcinogenic PAHs was observed in the control, and at the end of the incubation the extractable concentration of carcinogenic PAHs was 74 mg/kg (Figure 8.5). The apparent increase in carcino-

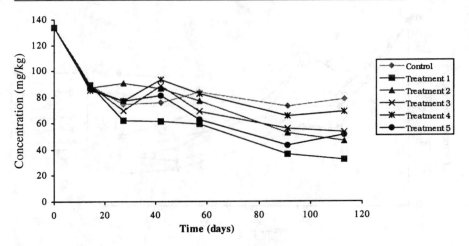

Figure 8.5. Influence of bioremediation treatments on carcinogenic PAH concentrations.

genic PAH concentration between day 27 and day 113 can possibly be attributed to variability in the spatial distribution of PAHs in the sediment.

Concentrations of carcinogenic PAHs were reduced more substantially in sediment subjected to bioremediation treatments. In sediment subjected to the most effective treatment, number one, carcinogenic PAHs were reduced by 54%, from 134 mg/kg to 57 mg/kg, in the first 27 days of incubation. During the remainder of the investigation, from day 27 to day 113, the carcinogenic PAH concentration was further reduced to 29 mg/kg (Figure 8.5), representing a cumulative reduction of 77% by the end of the incubation. Other bioremediation treatments resulted in less extensive reductions in carcinogenic PAH concentrations. The results indicate that appropriate bioremediation treatment can be used to reduce the concentrations of carcinogenic PAHs in dredged harbor sediments to below the regulatory criteria for many jurisdictions.

PAH Degradation Rates

Degradation rates for individual PAH compounds in sediment subjected to bioremediation treatment one were calculated assuming first-order degradation kinetics. The first-order rate coefficient, predicted initial concentration, first-order reaction rate constant, and half-life for each PAH are presented in Table 8.4. With the exception of acenaphthylene, which had a low initial concentration and tended to fluctuate in concentration throughout the incubation, the half-life determined for each PAH was 155 days or less. Compared to published data on

Table 8.4. First-Order Rate Coefficient, Predicted Initial Concentration, Initial Concentration and Estimated Half-Life for Selected PAHs in Sediment Treated by Treatment.

Parameter	First-Order[a] Rate Coeff. (1/day)	Predicted Initial (mg/kg)	Measured Initial (mg/kg)	Half-Life (days)	Reported[b] Half-Life (days)
Naphthalene	0.0120	34	74	58	16.6 – 48
Acenaphthylene	-0.00631	0	0.81	-110	42.5 – 60
Acenaphthene	0.0198	9	35	35	12.3 – 102
Fluorene	0.0175	11	38	40	32 – 60
Phenanthrene	0.0179	33	111	39	16 – 200
Anthracene	0.0119	4	18	58	
Fluoranthene	0.0244	74	86	28	140 – 440
Pyrene	0.0227	63	61	31	210 – 1,898
Benzo(a)Anthracene	0.0168	22	30	41	102 – 679
Chrysene	0.0161	17	22	43	372 – 993
Benzo(b)Fluoranthene	0.0107	30	34	65	360 – 610
Benzo(k)Fluoranthene	0.0118	9	9.2	59	909 – 2,139
Benzo(a)Pyrene	0.0102	15	22	68	57 – 529
Indeno(1,2,3-c,d)Pyrene	0.00520	11	12	133	599 – 730
Dibenz(a,h)Anthracene	0.00640	3	4.0	108	361 – 942
Benzo(g,h,i)Perylene	0.00448	12	14	155	590 – 650

[a] Rate constant, predicted initial concentration, and half-life were derived assuming first-order degradation kinetics.

[b] From Howard, P.H., R.S. Boethling, W.F. Jarvis, W.M. Meylan, and E.M. Michalenko, *Handbook of Environmental Degradation Rates*, Lewis Publishers, Boca Raton, FL, 1991.

PAH half-lives under aerobic conditions in soil, the half-lives determined for PAHs in sediment in this study compare very favorably, especially for the HMW and carcinogenic compounds (e.g., the half-life for benzo[b]fluoranthene in sediment subjected to Treatment 1 was 65 days compared to a natural half-life of 360 to 610 days). The distinction was less clear for the lighter, more volatile PAHs.

Non-PAH Parameters

The influence of bioremediation treatments on a variety of non-PAH parameters was also assessed. The results from analyses conducted on the air-dried control sediment and sediment subjected to the two most effective treatments, number 1 and number 5, are presented in Table 8.5.

Bioremediation treatments had no apparent influence on the concentrations of metals, which were below relevant criteria both before and after the 113-day incubation. The absence of any substantial change in metals concentrations suggests that dilution was not a factor in the large reduction observed in PAH concentrations. The results led to the conclusion that most of the organic amendment added, a total of 8% wt/wt for treatment number 1, was mineralized during the 113-day incubation. This interpretation is consistent with studies on turnover of plant materials in agricultural soils.[24]

The apparent decrease in lead concentration cannot be explained. The decrease cannot be due to enhanced binding of metals to the added organic matter (i.e., the DARAMEND organic amendment), since the same reduction was observed in the air-dried control sediment which was not amended. Following 113 days of maintenance, the concentrations of every measured chlorophenol species was below the analytical detection limits, which ranged from approximately 0.1 mg/kg to 0.8 mg/kg.

The dioxin and furan TEQ in sediment subjected to the most effective bioremediation treatment, number 1, fell by 24% from 251 ng/kg to 190 ng/kg during the 113-day incubation. It is, however, impossible to determine if the change is significant because the endpoint values are from one sample and therefore statistical analyses cannot be conducted. As noted for the metals, above, the decrease cannot be attributed to dilution by organic amendment addition.

Bioremediation treatments significantly reduced the oil and grease content of the sediment.

Toxicity Data

The results of toxicity bioassays conducted on the untreated sediment and the sediment subjected to the most effective treatment indicated that

Table 8.5. Concentrations of Selected Non-PAH Parameters Initially and Following 113 Days of Maintenance Compared with CCME and MOEE Criteria.

Parameter	Units	CCME Criterion[a]	Initial	Air-Dry Control	Treatment 1	Treatment 2
Aluminum	mg/kg	—	20,400	15,600	16,200	17,900
Arsenic	mg/kg	50	<3.4	<3.5	<4.9	<4.7
Cadmium	mg/kg	20	<0.2	<0.2	<0.3	<0.3
Chromium	mg/kg	800	37	34	34	36
Copper	mg/kg	500	43	45	45	48
Iron	mg/kg	—	30,500	28,100	28,500	30,800
Lead	mg/kg	1,000	14	<2.0	<2.8	<2.7
Manganese	mg/kg	—	313	n.a.[b]	n.a.	n.a.
Nickel	mg/kg	500	27	25	25	27
Zinc	mg/kg	1,500	98	99	98	103
Pentachlorophenol	mg/kg	5	0.107	<0.101	<0.101	<0.101
Oil and grease	mg/kg	—	35,100	39,400	21,800	17,000
Dioxins and furans TEQ	ng/kg	1,000[a]	251	271	190	217

[a] Interim guideline for industrial land use, established by Council of Canadian Ministers of Environment. Water Quality Branch, Environment Canada, Ottawa, Ontario, 1991.

[b] Not analyzed.

bioremediation reduced or eliminated sediment toxicity. After 5 days there was no significant difference in the germination of timothy grass or leaf lettuce in the control and treated sediment. Seedlings that germinated in the sediment subjected to bioremediation treatment were, however, larger and more robust than those that germinated in the control sediment.

All earthworms released into the control sediment died within 7 days, while all earthworms placed in the treated sediment remained alive and showed no visible symptoms of stress during the 14-day test period.

These data suggest that the sediment subjected to bioremediation treatment was less toxic and therefore the sediment may be suitable for more uses following remediation.

PILOT-SCALE BIOREMEDIATION DEMONSTRATION

A pilot-scale demonstration of sediment bioremediation, in which the same bioremediation treatment as that described for the bench-scale study was applied to approximately 100 tons of dredged sediment, was conducted at Hamilton Harbour, near Burlington, Ontario. The results of the demonstration, which were published earlier,[11] mirror those obtained in the bench-scale study described here. Briefly, the total PAH concentration was reduced by 90%, from approximately 1,139 mg/kg to approximately 110 mg/kg, during 10 months of treatment. Concurrently, the concentrations of high molecular weight and carcinogenic PAHs fell from 293 mg/kg and 155 mg/kg to 82 mg/kg and 48 mg/kg, representing reductions of 72% and 69%, respectively. Results of radioisotope fate studies similar to those described above for the bench-scale studies, indicated that the mechanism responsible for the observed reductions in PAH concentrations was complete microbial biodegradation of the compounds with release of PAH carbon as carbon dioxide. Tomato seed germination studies conducted on the untreated control sediment and the sediment following 10 months of bioremediation indicated that bioremediation sharply reduced the phytotoxicity of the sediment. Specifically, the bioremediated sediment supported 68% germination of tomato seeds, while in the untreated control sediment germination was restricted to 10%.

SUMMARY AND CONCLUSIONS

The results of bench-scale and pilot-scale studies conducted on PAH-impacted sediments dredged from harbors on Lake Superior and Lake Ontario indicated that appropriate bioremediation treatment can provide an effective means of decontamination. The studies also indicated that bioremediation treatment conditions can sharply influence the rate at which PAHs are biodegraded. Radioisotope fate studies indicated that bioremediation treatments can increase the rate and extent of biodegradation of PAHs to carbon dioxide. Seed germination and earthworm mortality studies indicated that bioremediation can substantially reduce sediment toxicity. When appropriate bioremediation treatments are employed, PAH concentrations can be rapidly reduced to regulatory criteria established for industrial land use in Canada.[25-27]

ACKNOWLEDGMENTS

A.G.S. acknowledges the support of Environment Canada through the Great Lakes Cleanup Fund, the Ontario Ministry of Environment, and the Site Remediation Division of Water Technology International Inc. J.T.T. acknowledges the support of the National Sciences and Engineering Research Council of Canada (NSERC) for supporting his research program. The skillful assistance of Ms. M. Sequeira in preparing the figures is also gratefully acknowledged.

REFERENCES

1. Smith, T.J., R.J. Wilcock, R.D. Pridmore, S.F. Thrush, A.G. Langdon, A.L. Wilkins, and G.L. Northcott. Persistence of chlordane applied to an intertidal sandflat. *Bull. Environ. Contam. Toxicol.*, 49, pp. 1535–1541, 1991.

2. Great Lakes Action Plan. *Proceedings: Workshop on the Removal and Treatment of Contaminated Sediments.* Environment Canada's Great Lakes Cleanup Fund, Managed by Wastewater Technology Centre, Burlington, Ontario, 1993.

3. Canadian Environmental Protection Act. *Priority Substances List Assessment Report: Creosote-impregnated Waste Materials.* Government of Canada, Environment Canada, Health Canada, Ottawa, Ontario, 1993.

4. *Sediment and Biological Assessment of the Northern Wood Preservers Inc. Site, Thunder Bay, July 1995 and September 1995, Final Report.* Environmental Monitoring and Reporting Branch and Standards Development Branch. Ontario Ministry of Environment and Energy, Toronto, Ontario, 1996.

5. *Bioremediation in the Field.* United States Environmental Protection Agency. Office of Research and Development, Cincinnati, OH, EPA/540/N-92/001, No. 5, pp. 11–31, 1992.

6. Mihelcic, J.R. and R.G. Luthy. Microbial degradation of acenaphthene and naphthalene under denitrification conditions in soil-water systems. *Appl. Environ. Microbiol.*, 54, pp. 1188–1198, 1988.

7. Berry, D.F., A.J. Francis, and J.-M. Bolag. Microbial metabolism of homocyclic and heterocyclic aromatic compounds under anaerobic conditions. *Microbiol. Rev.*, 51, pp. 43–59, 1987.

8. State of New Jersey. *Bid Documents: Sediment Decontamination Demonstration Project.* Department of the Treasury, Bid Number: 98-X-99999, Volume 1, 1998.

9. Dagley, S. Lessons from Biodegradation. *Ann. Rev. Microbiol.,* 41, pp. 1–23, 1987.

10. Menzie, C.A., B.B. Poticki, and J. Santodonato. Exposure to carcinogenic PAHs in the environment. *Environ. Sci. Technol.,* 26, pp. 1278–1284, 1992.

11. Bucens, P., A. Seech, and I. Marvan. Pilot-scale demonstration of DARAMEND bioremediation of sediment contaminated with polycyclic aromatic hydrocarbons in Hamilton Harbour. *Water Qual. Res. J. Can.,* 31, pp. 433–451, 1996.

12. Heitkamp, M.A. and C.E. Cerniglia. Mineralization of polycyclic aromatic hydrocarbons by a bacterium isolated from sediment below an oil field. *Appl. Environ. Microbiol.,* 54, pp. 1612–1614, 1988.

13. Cerniglia, C.E. and M.A. Heitkamp. Microbial Degradation of Polycyclic Aromatic Hydrocarbons (PAH) in the Aquatic Environment, in *Metabolism of Polycyclic Aromatic Hydrocarbons in the Aquatic Environment,* Varanasi, U., Ed., CRC Press, Inc., Boca Raton, FL, 1989.

14. Cerniglia, C.E. Biodegradation of polycyclic aromatic hydrocarbons. *Biodegradation,* 3, pp. 351–368, 1992.

15. Foght, J.M. and D.W.S. Westlake. Degradation of polycyclic aromatic hydrocarbons and aromatic heterocycles by a *Pseudomonas* species. *Can. J. Microbiol.,* 34, pp. 1135–1141, 1988.

16. Gardner, W.D., R.F. Lee, K.R. Tenore, and L.W. Smith. Degradation of selected polycyclic aromatic hydrocarbons in coastal sediments: importance of microbes and polychaete worms. *Water, Air Soil Pollut.,* 11, pp. 339–347, 1979.

17. Hazen, T.C., L. Jimenez, G.L. de Victoria, and C.B. Fliermans. Comparison of bacteria from deep subsurface sediment and adjacent groundwater. *Microbial Ecol.,* 22, pp. 293–304, 1991.

18. Dagley, S. Biodegradation and biotransformation of pesticides in the earth's carbon cycle. *Residue Rev.,* 85, pp. 127–137, 1983.

19. Ingraham, J.L., O. Maaloe, and F.C. Neidhardt. *Growth of the Bacterial Cell.* Sinauer Associates, Inc., Sunderland, MA, 1983, pp. 251–261.

20. Atlas, R.M. *Microbiology: Fundamentals and Applications.* MacMillan Publishing Company, New York, 1988, pp. 347–348.

21. Aquatic Ecosystems Objectives Committee. *Annual Report to the Great Lakes Science Advisory Board of the International Joint Commission.* IJC Regional Office, Windsor, Ontario, 1983.

22. Greene, J.C. *Protocols for Short Term Toxicity Screening of Hazardous Waste Sites.* U.S. EPA, 3-88-029, Corvallis, OR, 1989.

23. Organization for Economic Cooperation and Development. *Guidelines for Testing of Chemicals: 207E—Acute Toxicity Tests.* ISBN 92-64-1222-4, 1990.

24. deCatanzaro, J.B. and E.G. Beauchamp. The effect of some carbon substrates on denitrification rates and carbon utilisation in soil. *Biol. Fertil. Soils,* 1, pp. 183–187, 1985.

25. CCME. *Review and Recommendations for Canadian Interim Quality Criteria for Contaminated Sites.* Scientific Series No. 197, The National Contaminated Sites Remediation Program, Inland Waters Directorate, Water Quality Branch, Environment Canada. Ottawa, Ontario, 1991.

26. MOEE. *Guidelines for the Protection and Management of Aquatic Sediment Quality in Ontario.* Water Resources Branch, Ontario Ministry of the Environment, Toronto, Ontario, 1992.

27. MOEE. *Guideline for Use at Contaminated Sites in Ontario.* Standards Development Branch, Ministry of Environment and Energy, Toronto, Ontario, 1996.

Index

R

Radioactive carbon dioxide   188
Radioimmunoassay (RIA)   129, 131
Radioisotope fate studies   188, 200
Rainbow trout (*Oncorhynchus mykiss*)   13, 30, 37, 38, 54, 61, 62, 66, 81, 83, 86, 91, 102, 105, 107, 108, 110, 125, 130, 145
Rana temporaria   100
Rat   62
   microsomes   68
Rearing solution   37
Rec-A operon   59
Receptor binding assays   130
Receptor binding properties of 17β-estradiol   150
Receptor hormone complex   127
Receptor mediated processes   124, 152
Recombinant bacterial strain   59
Recombinant cell cultures   131, 135, 151
Redox potentials   185
Refinery effluent
   in situ exposure   148–150
Renewal protocol   144
Reporter gene   135
Reproductive abnormalities   29, 53, 80, 124, 126, 129
   alligators   124
Reproductive cycle   30, 116
Reproductive steroids   80
Reproductive success   30, 46
Residence time   83
Residual chlorine   162
Resin acids   91, 94
Resin adsorbents   57
Revertant cells   58, 59
Revertant colonies   63
Rotoevaporation   56, 57, 81

S

S-9 activation   71
S-9 microsomes   62, 103, 104
Saint-Lawrence River   176
Saint-Lawrence River Action Plan   162, 177
Salmonella typhimurium   53, 54
Schiff-positive microbodies   61
Schiff's reagent   61
Scintillation cocktail   62, 188
Scintillation counter   62
Screening   165
Secondary sex characteristics   34, 39, 41–44, 79–80, 125
Secondary treatment   79, 83, 91, 147
Sediment   184
   anaerobic   184
   bioremediation   200
   carcinogens   22
   contamination   1
   toxicity   184, 187, 190, 200, 201
   water interface   184, 185
Seed germination   190
Selenastrum capricornutum   163, 170
   assay   173
Seminiferous tubules   34
Serum steroid   30
Serum testosterone and estradiol   125
Sewage treatment   123, 161, 162, 173, 177
   effluents   129, 130
Sex determination   126
Sex differentiation   29, 38, 105–106
Sex hormones   30, 42
   ratios   150
Sex reversal   32, 34, 38–41
Sex steroids   30
Sexual bipotentiality   40